Analytical Aspects
of Drug Testing

CHEMICAL ANALYSIS

A SERIES OF MONOGRAPHS ON ANALYTICAL CHEMISTRY AND ITS APPLICATIONS

VOLUME 100

WILEY

A WILEY-INTERSCIENCE PUBLICATION

JOHN WILEY & SONS

New York / Chichester / Brisbane / Toronto / Singapore

Analytical Aspects of Drug Testing

Edited by

DALE G. DEUTSCH

State University of New York at Stony Brook
Stony Brook, New York

with foreword by
Irving Sunshine

WILEY

A WILEY-INTERSCIENCE PUBLICATION

JOHN WILEY & SONS

New York / **Chichester** / **Brisbane** / **Toronto** / **Singapore**

Library of Congress Cataloging in Publication Data:
Analytical aspects of drug testing / edited by Dale G. Deutsch.
     p. cm.—(Chemical analysis ; v. 100)
    "A Wiley-Interscience publication."
    Includes bibliographies.
    ISBN 0-471-85309-7
    1. Drugs—Analysis. 2. Drug testing. 3. Drug abuse—Prevention.
I. Deutsch, Dale G. II. Series.
RB56.A52 1988
616.86'307'56—dc19   88-17291
                                             CIP

Printed in the United States of America

10 9 8 7 6 5 4 3 2 1

CONTRIBUTORS

John C. Baenziger, Indiana University Medical Center, Indianapolis, Indiana

Dale G. Deutsch, State University of New York at Stony Brook, Stony Brook, New York

Kenneth Emancipator, State University of New York at Stony Brook, Stony Brook, New York

Martha R. Harkey, University of California, Davis, Davis, California

Dennis W. Hill, University of Connecticut, Storrs, Connecticut

Peter I. Jatlow, Yale University School of Medicine, New Haven, Connecticut

Richard W. Jenny, New York State Department of Health, Albany, New York

Jean C. Joseph, St. Mary Medical Center, Long Beach, California

Karen J. Langner, University of Connecticut, Storrs, Connecticut

Arthur J. McBay, University of North Carolina, Chapel Hill, North Carolina

R. H. Barry Sample, Indiana University Medical Center, Indianapolis, Indiana

Barbara A. Smith, St. Mary Medical Center, Long Beach, California

Robert D. Voyksner, Research Triangle Institute, Research Triangle Park, North Carolina

Steven H. Y. Wong, University of Connecticut School of Medicine, Farmington, Connecticut

FOREWORD

Analytical toxicologists are faced with ever-increasing demands for more sensitive and specific methods to detect the increasing number of clinical agents in common use. To help satisfy this demand, the editor of this volume has assembled a series of monographs describing the cutting edge of recent advances in analytical techniques. By adapting these techniques to the analysis of biological specimens, toxicologists will be able to provide data to those concerned that may help resolve problems related to proper use, misuse, or abuse of drugs. These problems include the ongoing search for: (a) causes of poisonings, (b) improved therapeutic approaches using currently available drugs, (c) the development of new drugs and their pharmacology, (d) means to insure a community, military, and industrial environment free of drug abuse, and (e) ascertaining the implications of substances detected in biological specimens during medicolegal investigations.

Analysts have responded to these demands in an exemplary fashion. They have developed techniques and procedures whose precision, accuracy, specificity, and sensitivity have met initial goals. As these goals were met, other more sensitive and specific demands were made because more knowledge and insights were acquired. This continuing process creates the need to modify the current procedures, a need that can be satisfied only by utilizing the latest modern technology.

The authors of the chapters in *Analytical Aspects of Drug Testing* provide significant information on these technological advances and suggest how they can be applied to drug analyses. Rightfully, the volume begins with the topic of quality assurance and ends with material designed to help interpret the results obtained by any of the innovative techniques described in between these two chapters. Also included in the volume are discussions on separation techniques, immunoassays, and the many chromatographic techniques that have become essential to adequate laboratory performance. In many instances these techniques involve relatively expensive equipment and specially trained personnel. The rising costs of health care will determine the extent to which these improved techniques will be used. The cost/benefit ratio will be a deciding factor in this decision.

Refined and improved laboratory data are achievable. Reliable as they may be, they may have little or no value in patient care if personnel beyond the laboratory fail in their responsibilities. *Cooperation* is the key word. The attending physician concerned with monitoring proper use of a therapeutic drug (a) must initiate an explicit request for a laboratory result; (b) must insure that the proper procedures are in place for obtaining the designated specimen at a proper time (it is essential that the exact time of sampling be noted as well as the time the last dose of medicine was given); (c) must provide for prompt delivery of the specimen to the laboratory; (d) should inform the laboratory of all relevant clinical information; and (e) should be available when the laboratory result is reported.

If a suspected poisoning is being investigated, both blood and urine specimens are required. A history of medications taken by the patient prior to the incident that led to the need for treatment, a list of medications administered to the patient by the attending staff, and the patient's presenting symptoms and current status should be reported to the toxicologist. In medico-legal investigations (and these include testing urines for drugs subject to abuse), a chain of custody must be established and maintained. This requires a written document listing all those who handle the specimens as they proceed from obtaining the specimen from the person involved in an incident to all the laboratory staff members doing the analyses. Laboratory security must be assured and written records must be kept of each person who handles each specimen, including the times of analyses, the storage and ultimate disposition of the specimens, and all original data on standards, quality control, and positive results. Criteria for maintenance, quality control and analytical procedures must be available in a suitable volume. Evidence of supervisory inspections of all results must be documented. Failure to observe these essential steps could vitiate the legal value of the most adequate laboratory result.

The laboratory, too, has responsibilities beyond that of providing an adequate method for a particular need. Quality assurance is essential to quality analyses. However accurate and reliable the analysis, the result will be undesirable and valueless if the sample is mislabeled, if it is confused with another, if proper records (and in medico-legal instances the related analytical data and the specimen) are not kept, if reports are not checked prior to release and then not reported properly. Quality assurance provides for continuous monitoring of all steps in the laboratory; it is essential. Too often, most laboratories consider quality control procedures (essential as they are) to be sufficient and equivalent to quality assurance. There is no substitute for quality assurance and it is obligatory in every laboratory.

Notwithstanding optimal compliance with all of the abovementioned details, a laboratory result is only of value if it is interpreted properly and put to use in patient care. Improper or inadequate interpretation of an excellent laboratory datum can negate all previous well-accepted procedures. Many clinical laboratory scientists feel their role is satisfied when they submit a proper report. This is debatable. They should put emphasis on the clinical aspect of the "clinical" laboratory scientist (or clinical chemist or clinical toxicologist) and collaborate with the clinician, when appropriate, to insure proper interpretation of a laboratory result. Clinical decisions should never be made solely on the basis of a laboratory result. This is all the more true in urinalysis for drugs subject to abuse, where misinterpretation of the laboratory data can result in an unwarranted discharge of employees.

To reiterate, there is more to analysis than the result of an executed analytical procedure. A significant amount of ancillary efforts are required, which involve laboratory personnel, clinicians, and administrators. It is essential that these groups be aware of their respective obligations and that they fulfill them in an exemplary way. This can only be achieved if communication pathways are patent and used as necessary. When this is accomplished, then laboratory data can be put to effective use in improved patient care.

IRVING SUNSHINE

Palo Alto, California

PREFACE

Drug testing affects many areas of our society. Among the most publicized are the workplace and employee programs, but it has also proliferated in athletic programs, the military, correctional facilities, and rehabilitation programs. Drug testing also plays a very crucial role in the medical setting. In the clinical laboratory the toxicologist performs therapeutic drug monitoring, emergency room toxicology for the overdosed or poisoned patient, and testing for substance abuse in patients from psychiatry, neurology, obstetrics and gynecology, pediatrics, sleep clinics, and other areas.

Due to the continued proliferation of new technology, the field of analytical toxicology is expanding vigorously, particularly in the area of drug testing. This book is designed to update readers on previously established technology, outline new techniques that are now being used in the more advanced laboratories and explore a number of techniques which are still in developmental stages. Many of the chapters in this volume contain data presented for the first time.

The first chapter is presented in recognition of the fact that sophisticated analytical technology does not, in itself, insure quality results. Chapter 1, written by Dr. Jenny, a scientist in charge of the New York State proficiency testing program, is a primer on quality assurance for laboratories performing clinical and workplace drug testing. It contains explicit procedures which, if followed, insure that the integrity of good analytical procedures is not compromised by any weakness in the steps in the process, from collection of the sample to the reporting of results. Chapter 1 includes such important topics as: specimen collection, documentation, the standard operating procedure manual (SOPM), laboratory security, quality control, and analytical method performance characteristics.

One of the simplest and most widely used methods for drug testing in the toxicology laboratory is the immunoassay. Assays using Syva EMIT reagents for each of ten drug classes are discussed in detail in Chapter 2. Comprehensive tables listing those drugs that yield a positive response as well as those which cross-react to produce a false positive are pre-

sented. Chapter 2 also includes the principles of the enzyme immunoassay, methods for specimen collection, assay conditions, and types of
commercial instruments used with the reagents.

As described in Chapter 3, solid phase extraction (SPE) procedures
are becoming popular in analytical toxicology because they are more
rapid, selective, and require smaller volumes of organic solvents than do
liquid–liquid extraction procedures. By applying the principles described
by the author of Chapter 3, Martha Harkey, one can develop an extraction
protocol for any drug. An appendix to Chapter 3 contains SPE methods
for many drugs, including delta-9-carboxy tetrahydrocannabinol, from
urine.

With the availability of less expensive, more compact, and easier-to-
use models of gas chromatograph/mass spectrometers (GC/MS), they are
becoming standard equipment for toxicology laboratories. In Chapter 4,
GC/MS for drug detection using library search routines is illustrated for
the identification of unknown substances as well as for the confirmation
of specific drugs. In Chapter 9, procedures for GC/MS screening and
analysis of anabolic steroids are described in detail. These state-of-the-
art methods were used in the anabolic steroid screening program during
the 10th Pan American Games in 1987.

The emergence of high performance liquid chromatography (HPLC)
for drug analysis is reflected by three novel chapters in this volume. Chapter 5 details methods of screening for drugs using HPLC with an ultraviolet
photodiode array detector. Drugs are identified by a combination of retention indices and ultraviolet spectra, using a computerized spectral library search program written by authors Hill and Langner. Two evolving
techniques for toxicology and therapeutic drug monitoring are microbore
liquid chromatography (MBLC) and direct-sample-analysis, both of which
are presented in Chapter 6. These methods fulfill the need to analyze
smaller volumes, such as those encountered in neonatal and pediatric
patient samples, with accelerated turnaround times. Examples are given
for the clinical analysis of theophylline, caffeine, procainamide, and chloramphenicol. In Chapter 7, a technique destined to become a routine tool
in toxicology combines the convenience of high performance liquid chromatography with the specificity of mass spectroscopy (HPLC/MS). Author Robert Voyksner describes applications of HPLC/MS for the analysis
of steroids, alkaloids, mycotoxins, drugs, and pesticides.

In some instances it would be desirable to perform toxicology tests
with dipsticks and observe color development visually or with a simple
machine—as is now done for glucose. Chapter 8 reviews the dry reagent
chemistries that have been developed for clinical toxicology and therapeutic drug monitoring. The principles of these various techniques are

described, with examples of theophylline, phenytoin, primidone, carbamazepine, gentamicin, tobramycin, phenobarbital, and amikacin.

No volume on drug testing would be complete if it did not, in addition to describing the analytical methodology, profile a drug of abuse. In Chapter 9 this is done for cocaine, a drug of abuse very much in the public eye. The clinical properties and toxicity, routes of administration, metabolic disposition and kinetics, neurochemical mechanism, and methods of analysis are presented. Furthermore, any volume in this area would be amiss if it did not, in addition to describing the analytical techniques, describe the problems and pitfalls of drug testing. The final chapter, Chapter 11, addresses these issues in terms of screening and confirmation tests such as the immunoassay and chromatographic methods, as well as discussing quality assurance issues.

Acknowledgment

I thank Steven H. Y. Wong for recommending me to Wiley to edit this volume when he was the chairman of the Therapeutic Drug Monitoring and Clinical Toxicology Division of the American Association for Clinical Chemistry. Lou C. Deutsch, my wife, generously provided me with grammar, punctuation, and emotional support throughout this undertaking.

DALE G. DEUTSCH

Stony Brook, New York
December, 1988

CONTENTS

CHAPTER

1

QUALITY ASSURANCE

RICHARD W. JENNY

New York State Department of Health
Albany, New York

Analytical toxicology, as performed in the clinical laboratory, is generally applied to two principal areas: (1) diagnosis and treatment of patients with acute or chronic exposure to toxic substances, and (2) substance abuse testing for treatment programs and the workplace. Laboratory activities required to conduct toxicological investigations include sample collection, transport, accessioning, analysis, and reporting of analytical results. Associated with each of these major activities are additional tasks that must be performed to ensure the integrity of the testing process.

1

Paramount among all laboratory activities, however, is the quality assurance program. Regardless of how meticulously an analytical procedure is performed, the usefulness of the testing result is compromised, perhaps negated, by any weakness in the test process chain. It is the laboratory's obligation to the clinician and the clinician's patients to establish guidelines for the proper execution of each testing component. The quality of each testing result cannot be "assured"; however, a quality assurance program, properly designed and implemented, minimizes the frequency and severity of testing errors.

The design of a quality assurance program must be a collaborative effort between the physician and the laboratory director. The physician's role in this process is to describe the laboratory service required for good patient care. The description must include the scope of toxicology service, desired assay performance characteristics (sensitivity, specificity, accuracy, precision), the anticipated impact on patient care, and the expected turnaround time for reports. With this information, the laboratory director can outline laboratory requirements for specimen type and volume, sample preservation and transportation, and documentation of pertinent clinical data that might guide the course of laboratory investigation and also permit interpretive reporting.

With the ground rules in place, the laboratory has the distinct advantage of making informed decisions in the development of a viable toxicology service. Objective guidelines for quality assurance can then be established to include the specific demands imposed by the health-care practitioner.

In this chapter, I first focus on the various applications of analytical toxicology services, describing quality assurance guidelines that are necessary if the laboratory is to produce a quality product that is consistent with medical requirements for good patient care. I then provide a more detailed discussion of the preparation and use of reference and quality-control materials and of the general elements of a quality assurance program.

1. CLINICAL TOXICOLOGY

Drug overdose and exposure to nondrug toxic agents are responsible for a sizable percentage of the total cases treated in our nation's emergency rooms (1). The symptoms produced by these agents mimic those that accompany many medical conditions; the urgency of the situation and the use of notoriously inaccurate clinical histories further complicate differential diagnosis (2). The toxicological analysis of body fluids is used to establish or exclude poisoning as the cause of the patient's acute illness.

The confirmation of suspected drug overdose or exposure to toxic substances permits discontinuation of further diagnostic steps and the application of specific therapeutic measures, if available.

1.1. Medical Requirements

Management of the poisoned patient generally involves support of vital functions until the poison is cleared from the system (3). Specific antidotal therapies, however, are available for a few intoxicants, for example, acetaminophen, salicylate, antidepressants, barbiturates, ethylene glycol, methanol, mercury, iron, and lead (3, 4). The antidotes are also potentially harmful to the patient and must be used judiciously. The intoxicant blood concentration is often of prognostic value and is used to assess the necessity for specific therapeutic intervention.

The clinical staff therefore requires a level of support that includes the following (5):

1. A broad-spectrum urine drug screen capable of detecting drugs commonly abused in the community it serves.
2. Quantitation or semiquantitation of the amount of drug in blood, when necessary and appropriate, to document the correlation of concentration and effect. This is done to rule out other underlying causes of the medical condition.
3. Reliable serial determinations of blood concentration of those drugs for which there is a specific antidote for assessment of prognosis and course of therapy.
4. Expeditious reporting of toxicological analyses to ensure prompt, optimal patient care.
5. Twenty-four-hour service.

1.2. Laboratory Requirements

Perhaps more than in any other area in laboratory medicine, the toxicologist requires the active participation of clinical staff during the toxicological investigation of any disease state of unknown origin. The toxicologist must rely upon the astute diagnostician to limit the universe of drugs to a drug or class of drugs that might produce the symptoms expressed by the patient. The laboratory's productivity and quality assurance in terms of patient care depend critically on the ongoing communication between clinical staff and the toxicologist.

Access to the patient during the acute crisis is generally limited to the medical staff, and the laboratory must rely upon the clinician to collect adequate volumes of the appropriate specimen for analysis. The laboratory is responsible for developing policy regarding specimen procurement, identification, and transport; however, the medical staff must exercise commitment in following the guidelines.

The toxicologist therefore requires a level of commitment on the part of the medical staff as follows (5, 6):

1. Criteria are established specifying when the laboratory service is to be initiated and the level of support required.
2. Clinical staff knows specimen procurement guidelines and makes every effort to conform.
3. Clinical staff provides as best it can all the information solicited in the test request form.
4. The physician maintains communication with the laboratory after the testing is initiated, informing the toxicologist of significant updates that might help the laboratory prioritize its work load.

1.3. Requisition Form

The test request form is the contract and principal means of communication between the toxicologist and the clinician. Properly designed, the requisition form provides an efficient mechanism for the clinical staff to provide pertinent information necessary for expeditious toxicological investigations. Patient identification and status, drug(s) or poison(s) suspected and approximate time of ingestion, drug(s) used in therapy, notation of specimen requirements for available toxicology services, specimens collected and time of collection, and the physician's name and location are important components of the requisition form. Provisions must be available to relate the specimen(s) to the requisition properly.

A tentative standard for the development of requisition forms for toxicology services has been issued (7) by the National Committee for Clinical Laboratory Standards (NCCLS). The proper design of the request form is essential for legibility and acceptance by medical staff. Henderson (8) has outlined factors to be considered in form design and provides useful examples of requisition forms used in his laboratory.

1.4. Specimen Considerations

There are many factors to be considered when selecting the appropriate specimen for analysis. The objective, however, should always be to generate the maximum amount of clinically useful information in the shortest time. To do so, the laboratory staff must be informed of how the toxicological analysis will affect patient management. The use of blood is appropriate when the substance concentration has prognostic value and/or will be used in the selection of treatment modalities. The use of urine is appropriate when detection of the substance in blood is difficult or when the etiology of the case is unknown and a broad-spectrum drug screen is required. Toxicologists have proposed varying themes on the best approach to the investigation of poisoning (9–11); sampling of blood, urine, or gastric content or washings (when available) may be optimal for expedient toxicological analysis.

Provided specimens are assayed soon after collection, the use of preservatives is not generally necessary (12); if the analysis is to be delayed several hours, samples should be refrigerated. If a preservative or anticoagulant must be used, the laboratory should first evaluate compatibility with analytical methods in use. For instance, β-lactam antibiotics (13) and heparin (14) have been reported to alter the reactivity of aminoglycosides in some immunoassays, and azide (15) produces a chromophore with spectral properties similar to that of the salicylate–ferric ion complex formed in the Trinder assay. Manufacturers of analytical kits are required to identify those preservatives and anticoagulants that will interfere in the determination of analyte; manufacturer recommendations on the selection of specimen treatment agents must be addressed.

The *Clinical Laboratory Handbook for Patient Preparation & Specimen Handling—Therapeutic Drug Monitoring/Toxicology* (Fascicle IV), published by the College of American Pathologists (12), provides a compilation of recommendations and literature references, when available, on specimen collection and handling techniques. For each of the drugs, metals, and toxins included in the compendium, the committee has outlined guidelines for patient preparation; preferred type of specimen, containers, and preservative(s) and anticoagulant(s); and storage requirements for the original and processed specimen. This publication is an excellent supplement to ongoing literature reviews and internal investigations of requirements for patient preparation and specimen handling procedures.

1.5. Analytical Method Performance Characteristics

The extent of emergency toxicology service is a function of the available expertise and resources. The laboratory must fully appreciate its limitations and offer only those analyses for which it has documented proficiency. Small community hospital laboratories can use relatively uncomplicated analytical methods and still provide a useful service. When it is considered that as many as 80% of the emergency room drug mentions tabulated by the Drug Abuse Warning Network involve salicylate, acetaminophen, opiates, cocaine, phencyclidine, ethanol, benzodiazepines, and tricyclic antidepressants (16), the impact of even the most rudimentary of services can be significant.

The principal methods of drug analysis are spot tests, immunoassays (RIA, EMIT, FPIA), thin-layer chromatography, gas chromatography, and liquid chromatography. Each has its limitations in scope, sensitivity, and specificity; chromatographic techniques require experienced technologists with the necessary interpretive skills. Method selection therefore should be preceded by a detailed assessment of laboratory capabilities and goals in terms of clinical need, a thorough review of the literature for descriptions of the performance characteristics and limitations of the method(s) being considered, and consultation with peers currently using the analytical system(s) in question.

After the method has been selected, the laboratory should verify manufacturer or literature claims of assay accuracy, recovery, precision, sensitivity, specificity, and linearity (17, 18). A reasonable amount of time should be allowed for the method evaluation, first to gain familiarity with the assay protocol and then to accumulate a data base of sufficient size for reliable assessment of performance characteristics. Because the gas and liquid chromatography methods are technically complex and the interpretation of thin-layer plates is often subjective, the competence of each analyst must also be evaluated if chromatographic methods are being considered.

The use of "split samples" to compare analytical results with those of a respected peer laboratory should be an integral part of the method evaluation; should discrepancies in results occur, the ensuing discussion with peers would be very educational. The process of comparing a laboratory's performance with that of peers is perhaps best achieved through participation in a toxicology proficiency testing program. Survey-validated specimens, available from testing program sponsors such as the College of American Pathologists, the American Association for Clinical Chemistry, and state departments of health that offer the service, provide an additional means of evaluating laboratory performance.

Perhaps the most difficult aspect of method evaluation is the determination of assay specificity, largely due to the vast number of substances that must be considered. Specificity studies generally entail the incorporation of compounds with chemical properties similar to those of the compound of interest into the appropriate matrix. The behavior of the potentially interfering substance in the assay is evaluated and documented by the analyst. Regardless of whether or not the substance interferes, it is essential that laboratory records include the statement that the compound was investigated. The record should include substance identification, concentration used, and assay characteristics, for example, the HPLC or GLC retention time, TLC R_f and reactivity to indicator reagents, color produced in spot tests and colorimetric assays, or relative reactivity in immunoassays. If the substance does interfere, a notation of its magnitude and significance should be provided and, as appropriate, alternative methods for differentiation specified.

The nature of assay interference is best characterized in-house. The danger of accepting manufacturer specifications of assay specificity is exemplified by a study of salicylate interference in a colorimetric acetaminophen procedure (19). Laboratories participating in a proficiency testing program were challenged with a set of five specimens containing acetaminophen and salicylate. The samples were labeled 4, 8, 12, 16, and 24 hr and the laboratory directors were informed that the samples simulated an overdose of both salicylate and acetaminophen. Laboratories were requested to report salicylate and acetaminophen concentrations, correcting the latter for salicylate interference if necessary. Eleven laboratories used a nitration procedure for acetaminophen determination known to suffer from salicylate interference; two of these did not correct for the interference and six used the correction factor supplied by the kit manufacturer. It was determined that the correction factor was incorrect; application of the laboratory results to the Rumack–Matthew nomogram, used by physicians to assess the need for therapeutic intervention (20), falsely indicated impending hepatic necrosis and the need for an intensive 3-day course of N-acetylcysteine therapy.

The laboratory should never assume that all interferents in its analytical systems have been identified. Investigations of assay specificity must be ongoing and should include literature reviews. Arrangements may also be made with the hospital pharmacist to list the drugs administered to those patients from whom blood and urine specimens are available to the laboratory. The specimens are then characterized as to drug and/or metabolite content, and hence are valuable for method evaluation and as a teaching aid.

The fact that assay specificity cannot be fully characterized addresses the need for confirmation of presumptive positive drug screens. The method used for confirmation should be of differing analytical principle from that used in the primary screen. Because confirmation testing usually does not take place until after preliminary findings have been reported to the emergency room physician, the physician is in a position to dictate the need and urgency for confirmation analysis. Regardless of the clinical need, the laboratory is obligated to assure the accuracy of its testing results and should perform confirmation testing if for no other reason than to evaluate assay methods.

When confirmation testing or a request for toxicological analysis is beyond the capability of the laboratory, it is necessary to use the services of a reference laboratory. Factors that must be considered when selecting a reference laboratory include laboratory credentials, accreditation or licensure, assay methods used, report timeliness, and cost. If the reference laboratory is being used to confirm in-house results, the referral laboratory must be certain that the confirmation method does not suffer the same limitations as that used for the primary screen, for example, an RIA procedure being used to confirm an EMIT result. Ultimately, the referral laboratory is responsible for the quality of testing results and should routinely monitor reference laboratory performance. This is best achieved by periodically submitting specimens of known drug content as patient samples. Should an erroneous report be received, the error source, the accuracy of previously submitted testing reports, and the potential adverse effect on patient care must be investigated.

1.6.   Quality Control

Once a method has been adopted, a quality-control program is implemented to monitor its long-term reliability. To do this, the laboratory must have access to a suitable supply of quality-control materials (see Section 4). Unlike routine therapeutic drug monitoring or clinical chemistry services, where specimens can be batched for purposes of cost effectiveness and scheduling, the emergency toxicology laboratory must be available on demand to provide toxicological services tailored to the needs of a given patient. As a general rule, the full complement of standards or calibrators, as appropriate, and controls must be assayed in parallel with the patient specimen. This requirement is changing, however, as technology produces analytical systems with exceptional calibration and reagent stability.

In fact, it is not possible to prescribe quality-control guidelines that would universally apply to assay methods. Each analytical method must be evaluated on its own merit as to the quality-control scheme required to monitor its performance. The nature of the control material, the number of levels, and the frequency of their analysis are dictated by the method performance characteristics and by the clinical application of the test.

Thin-layer chromatographic procedures require the grouping of the patient specimen(s) with the appropriate standard or standards and a positive and negative control. The standard is generally an organic solvent containing each drug included in the screen. The control material contains the drugs of interest in a matrix similar to that of the patient specimen. Ideally, a control specimen should be available for each drug/metabolite screened; this may be impractical, however, for laboratories offering a comprehensive screen. The use of a drug that exhibits extraction efficiency and reactivity to detection reagents that is characteristic of other drugs in its pharmacological class may be adequate. The control specimens must be carried through the entire procedure in parallel with the patient sample to verify the reliability of the testing process. Because the standards can be used to make judgments on the acceptability of the assay in terms of plate development and detection reagents, the control is used principally to monitor the reliability of the sample preparation stage. By preparing a control to contain screened drugs at a concentration approximating the respective limits of detection, analysis of the control serves the dual purpose of monitoring extraction efficiency and assay sensitivity.

Spot tests, although considered easy to perform, must also be adequately standardized and controlled. Because test reagents are applied directly to an aliquot of specimen without the need for drug extraction, a specimen containing the drug in question at a concentration approximating the assay detection limit can be used both as the standard and as the control. A blank control must also be assayed to ascertain the appearance of a "negative" sample and to check for possible contamination of assay materials. It is essential that the control materials be of the same matrix as the patient specimen.

When immunoassay is used as the primary screening method, the use of a positive and negative control in addition to the calibrator and patient specimen for each drug included in the screen may be cost-prohibitive. Wilson (21) has proposed a useful, cost-effective alternative for the quality control of immunoassays (specifically the EMIT-st) that assures assay reliability. By his scheme, each new lot of reagents is evaluated with positive and negative control specimens to ensure that assay parameters are within specifications. When a drug screen is requested, a calibrator and patient sample are assayed for each drug in the panel; the positive and negative control are assayed for a single, preselected drug. With each subsequent screen, the process is repeated with the exception of assaying positive and negative controls for the drug next in a rotation process. It must be stressed, however, that in those instances where a patient specimen tests positive for a drug, the assay must be repeated using positive and negative controls for the drug in question. If immunoassay is being used to confirm a presumptive positive screening result produced by another method such as TLC, positive and negative controls must routinely be included in confirmation testing.

Quantitative assays require the use of a minimum of two controls to monitor adequately assay accuracy and precision in the clinically relevant nontoxic and toxic ranges of substance concentration (17). The range of acceptable control results, criteria for quality-control data interpretation, and guidelines to be followed when an assay is deemed to be "out of control" must be documented and readily accessible to the analyst (see Section 5). In addition to recording quality-control data, the analyst should routinely record assay parameters (e.g., chromatography retention times, peak symmetry, column efficiency) that are an index of how well a method is performing. Such records, in concert with quality-control data, will allow a thorough assessment of error cause should an analytical problem arise.

Upon completion of the testing cycle, the laboratory supervisor should review the analyst's testing results and quality-control data, certifying report accuracy before it is released.

1.7.   Reporting

The urgency of the analysis often necessitates oral communication of testing results to the treating physician. This mode of communication can be beneficial to both parties because it offers the opportunity for timely exchange of information. The toxicologist can relate the analyses performed, results obtained, and the status of confirmation testing; if quantitative data are available, information relating concentration to clinical condition can be discussed. In light of the information provided by the laboratory, the physician may request additional testing.

Oral reports must always be followed by a written laboratory report. The report should include, in addition to pertinent clinical data, a list of specimens analyzed, each drug screened or quantitated, and the testing results (10). Because the drugs are listed, a negative report is a specific finding and does not mislead the physician as to the scope of the testing performed. A negative result should be referenced to the assay limit of detection and is best reported as "not detected." If a positive result has not been confirmed, the report should clearly state the result as "unconfirmed" positive and list the potentially interfering substances.

2.   SUBSTANCE ABUSE TESTING

Urine screening for drugs of abuse has become a major activity of many analytical toxicology laboratories. Rehabilitation programs, correctional facilities, government agencies, private industry, athletic commissions, and the educational system are increasingly relying on the laboratory to document the effectiveness of policies established to control and eliminate drug abuse (22). An individual's failure to comply with policy as evidenced by a positive drug test often results in punitive action that may vary from loss of earned "good time" for an inmate of a correctional facility, to loss of job or elimination from athletic competition. Failure on the part of the laboratory to detect drug abuse may give a false sense of security to those responsible for public safety, compromise the effectiveness of rehabilitation programs, and unduly delay counseling, treatment, and rehabilitation.

Clearly, the laboratory has an important role in the identification of the drug abuser; however, as in other disciplines of laboratory medicine, the toxicologist must have the support of a physician who is committed to the success of the program and to individual rights. Interpretation of a drug testing report in light of medical history and other pertinent information (e.g., employee records, treatment program progress reports) is essential when investigating the suspicion of drug abuse. A dialogue between the medical officer and the toxicologist is appropriate in those instances when testing results conflict with clinical impression: information of an individual's documented prescription or over-the-counter drug use may help the toxicologist resolve equivocal drug analyses; a physician's inquiry may uncover a laboratory administrative error. This collaborative effort is essential if the goals and objectives of drug detection and treatment programs are to be realized. Because drug testing results can end in punitive action toward an individual, it is incumbent upon the laboratory to produce definitive data. Protection of an individual's due process rights requires the laboratory to institute procedures that ensure the integrity of the sample and to conduct all aspects of the testing process with *documented* reliability. The relatively high potential of legal challenge therefore requires the toxicologist to augment quality assurance practices established for other areas of laboratory testing; these are discussed below.

2.1. Laboratory and Client Communication

The analytical approach to the processing of test specimens depends on the intended use of test findings. It is essential, therefore, that the laboratory be properly informed of how the information it produces is to be used. Hospital-affiliated laboratories that provide support to the emergency department establish a working relationship with treating physicians; hence requirements for toxicological analysis are established on an ongoing, collaborative basis. Laboratories removed from the physician, on the other hand, generally must act solely upon a requisition for laboratory analysis.

Decisions regarding the nature of laboratory service are predicated on the exchange of relevant information between the physician and the laboratory. This exchange is generally accomplished by use of a test requisition form. Clearly, test requirements (e.g., chain of custody, assay sensitivity, and specificity) for an employee drug testing program differ from those of a treatment program, which in turn differ from the needs of correctional facilities and probation departments. The requisition form therefore should clearly define the levels of toxicology service offered and state, or provide reference to, descriptions of specimen collection and handling requirements. Knowledge of the intended use of test findings will allow the laboratory to determine if the specimen(s) have been properly collected and handled.

The laboratory should also provide educational materials and/or on-site instruction in the proper use of laboratory services. This is particularly important in the workplace where personnel are not familiar with the precautions that must be exercised in specimen collection and handling. If conditions exist that allow specimen adulteration or switching and incomplete or inaccurate documentation of specimen identification, then the efforts put forth by the laboratory to ensure the integrity of the test process are nonproductive and the objectivity of the drug testing program is lost. An open line of communication between the laboratory and its client should be in place to address specimen collection and handling concerns.

Often the employer must use the services of a local laboratory for specimen collection and referral to the testing laboratory. Under such an arrangement, the testing laboratory should inform itself of the objectives and policies of the employer's drug testing program. The testing laboratory will then be able to select the appropriate test procedures and to inform the collection site personnel of proper specimen collection and handling practices.

2.2. Specimen Collection and Handling

The laboratory director must instruct his clients on proper specimen collection and handling procedures that are required to ensure specimen integrity. Variables that must be addressed include facilities used for collection, patient preparation, specimen type and volume, use of preservatives, documentation of patient identity, and if necessary, specimen chain of custody, specimen storage conditions, and specimen transportation to the testing laboratory (23).

Urine is the preferred specimen for substance abuse screening: collection is noninvasive, an ample amount is readily available, and detection of drugs and/or metabolites is possible for a longer period than in blood. In situations of "probable cause," blood samples should also be collected as soon after the incident as possible to address the question of whether or not a sufficient amount of drug was present to cause impairment.

The relatively few studies on the stability of drugs in body fluids have shown that the common test analytes, that is, amphetamines, barbiturates, opiates, benzoylecgonine, phencyclidine, benzodiazepines, and cannabinoids, are quite stable (12, 24, 25). These drugs in body fluids collected without preservatives and stored at 2–8°C are stable for weeks, and at − 20°C, for months (12). Cocaine in blood or plasma is rapidly metabolized by esterases but can be stabilized for up to 8 days by use of 0.5% w/v sodium fluoride and storage at 4°C (26). Cocaine is also unstable in alkaline urine but is stable for weeks at pH 5 (27). Special precautions are also necessary for collection of ethanol specimens. These include the use of nonalcohol wipes to clean the venipuncture site, the preparation of specimen to contain at least 0.5% w/v sodium fluoride to inhibit growth of ethanol-producing organisms, and filling the collection container to at least 90% of its capacity (12, 28). Specimens stored frozen are stable for months (28).

Policies and procedures for specimen collection and handling must address (1) the likelihood that a drug abuser will attempt to avoid detection and (2) the concern of program participants that they will be falsely identified as abusing drugs. The formulation of policies and procedures should be a collaborative effort between the laboratory director and his client and must address the concerns and needs of both parties. The laboratory director generally does not provide routine supervision of collection site personnel; however, if test specimens or documentation do not conform to prearranged standards, the director is obligated to provide feedback to his client.

The collection site should be a secure facility, containing all supplies and equipment necessary for specimen collection, storage, and shipping to the testing laboratory. Arrangements should be provided that allow privacy during urination. Bluing agents should be added to toilet water reservoirs and other sources of water should be removed to deter specimen dilution; soap dispensers and cleaning agents should be removed to deter specimen adulteration. The facility must be staffed by trained personnel who have been provided with a standard operating procedure manual that details the program's updated procedures for specimen collection and handling.

Patient preparation and collection procedures are designed to deter specimen "switching" or adulteration. The procedures must strike a balance between allowing privacy for the individual and instituting constraints that are necessary to ensure the integrity of the specimen. Guidelines prepared by the Department of Health and Human Services for federal agencies (29) are a suitable model for most drug testing programs.

Immediately upon receipt of the specimen, the collection supervisor should inspect the sample for evidence of adulteration or switching. Specimen temperature, color, odor, pH, and specific gravity should be recorded on the laboratory requisition or chain-of-custody form. If any of the specimen characteristics exceed predetermined criteria or limits of acceptability, the testing program policy and procedures for repeat collection must be implemented.

After the specimen has been submitted to the collection site person, it is important that the packaging of the specimen be conducted in open view of the individual being tested. Each container used for collection should have a label affixed that contains the following information: facility name, identification of the donor (e.g., name, social security number, or employee ID number), date and time of collection, name of the collection supervisor, and approximate specimen volume (23). The individual being tested should then initial the label, thereby acknowledging that the specimen is his/hers and is properly labeled.

Some applications of analytical toxicology, for example, employee drug testing, require that forensic principles be applied to specimen collection and handling. Procedures presented above apply; however, it is also necessary that tamperproof seals be applied to specimen containers and that all individuals involved in the collection, handling, and shipping of the specimen record their involvement on a chain-of-custody document. The seal is placed over the cap and down the sides of the container. The collection date and time and the initials of both the collection supervisor and the individual being tested are recorded on the tamperproof seal as evidence that the seal was placed on the specimen container by the collection supervisor in the presence of the specimen donor.

All specimens should be securely packaged along with test requisition/ custody forms and placed into secure storage, preferably a secure refrigerator, until shipment. If specimens are batched before they are transported to the testing laboratory, an invoice should be prepared to allow verification by the laboratory that all specimens have been received.

2.3. Laboratory Security

Laboratories have the legal responsibility to ensure the integrity and accurate identification of all test specimens regardless of the type or nature of the analysis. The forensic nature of substance abuse testing, however, requires additional documentation and security measures if laboratory analyses are to be legally defensible.

Access to those sections of the laboratory where specimens and records are stored and where analytical work is performed must be limited to authorized personnel. The laboratory director must maintain a log of individuals authorized to access these areas. Each time the secure areas are accessed, the individual must document his identity as well as the time, date, and purpose. Unauthorized individuals with a legitimate reason for seeking access must be escorted by authorized personnel.

In effect, directors who provide clinical and forensic services need to maintain two functionally independent laboratories. If computer resources for specimen accessioning, report generation, and billing are shared, the confidentiality of drug testing records must be ensured. As specimens for forensic analysis are received, they should be segregated from the clinical work and forwarded to the appropriate secure section of the laboratory.

2.4. Documentation

The laboratory chain of specimen custody begins with the examination of shipping and specimen seals. The invoice, if supplied, is reviewed to verify that all specimens have been received. The individual responsible for logging samples records his/her name, date and time, specimen condition, laboratory accession number, and integrity of container seals. This information is recorded on the chain-of-custody form that accompanies the specimen(s). If the condition or integrity of the specimen is suspect, analysis should not be performed and the laboratory should promptly inform its client of the situation.

When a specimen is accepted for analysis, it is incumbent upon the laboratory to document fully each aspect of the testing process. Chain-of-custody documentation must be maintained for both the original specimen and aliquots taken for processing. This documentation supports the presumption that (1) the integrity of the original specimen has been maintained if the need for retesting arises and (2) the reported test findings were derived from the analysis of authentic aliquots. All records of analysis, for example, work sheets, chromatograms, analyzer records, and reports, must be clearly labeled and traceable to the specimen analyzed and to the test requisition. Should the test results be challenged, documentation must be sufficiently complete to permit a reconstruction of the entire testing process. Calibration and quality-control data generated at the time of sample analysis must also be available to document test reliability.

2.5. Analytical Method Performance Characteristics

Methods used for substance abuse testing are essentially those used for detection of drug overdose. Enzyme immunoassay, radioimmunoassay, and thin-layer chromatography techniques are amenable to screening large batches of test specimens: the urine drug screen is performed to eliminate from further consideration the majority of the specimens that are drug-free. All presumptive positive drug screens must then be confirmed by an analytical method based on chemical and physical principles that differ from those used in the screening procedure. Methods used for screening or confirmation testing must be recognized as reliable by experts in the field of analytical toxicology.

Gas chromatography/mass spectrometry is widely accepted as the definitive method for confirmation testing (30). With the introduction of mass selective and ion trap detectors, most toxicology laboratories can now afford to purchase GC/MS instruments. Before the technology is implemented for confirmation testing, the laboratory director must verify that analysts are properly trained and that they possess the interpretative skills required for reliable specimen analysis.

It is not possible to equate the presence or amount of drug in urine with level of impairment; testing is performed simply to detect prior use of a drug. The laboratory therefore should optimize the sensitivity of its analytical methods to permit detection of drug/metabolite(s) at concentrations associated with casual, recent use. The judicious selection and use of analytical methods require the analyst to know the performance characteristics of methods and to understand drug pharmacokinetics. As an example, heroin is rapidly metabolized to morphine, and morphine in turn is excreted into urine principally as morphine-3-glucuronide (31). Immunoassay techniques are sensitive to the presence of morphine and its glucuronide conjugates; however, chromatographic techniques detect only free morphine. Therefore, morphine screening or confirmation of a presumptive positive immunoassay screen by a chromatographic procedure should be performed after conjugate hydrolysis (32). When class-specific immunoassay techniques are used, such as the EMIT amphetamine, benzodiazepine, or opiate assay, the analyst needs to be aware of the relative cross-reactivity and detection limit for each specific drug. The recently introduced benzodiazepine triazolam, for instance, must be taken in overdose to produce a positive result by the EMIT benzodiazepine assay (33).

The documentation of assay performance characteristics is also required to establish detection limits and cutoff values for screening and confirmation procedures. The detection or sensitivity limit of a method is the lowest concentration of drug in a specimen that can be detected reliably, generally at the 95% confidence level. As the drug concentration approaches the assay method's limit of sensitivity, it is increasingly difficult to discriminate between the analytical noise and the response produced by the drug; therefore, the use of the assay minimum detectable concentration to determine the presence or absence of drug is associated with a significant risk of false positive results.

To minimize the risk of false positive results, a cutoff concentration higher than the detection limit is selected; only those specimens producing an analytical response equivalent to or greater than the response produced by the cutoff calibrator (standard) are considered positive for the drug in question. Several criteria must be fulfilled when selecting an appropriate cutoff value: (1) the value should be sufficiently low to permit detection of recent, casual drug use; (2) the value should be sufficiently high to rule out interference by analytical noise or variation and by specimen matrix effects; (3) the cutoff level of the confirmation procedure should be equivalent to or lower than that of screening procedures to minimize the number of unconfirmed presumptive positive tests; and (4) when necessary, the cutoff level should be established to eliminate positive test results that are due to inadvertent exposure to drug, such as passive inhalation of tetrahydrocannabinol.

There are limitations inherent in the use of cutoff levels for labeling a urine test result positive or negative. Owing to analytical variability, determination of the status of drug presence or absence is not absolute. If urine drug concentration is equivalent to the cutoff level, about 50% of the drug test results will be falsely reported as negative. At increased (or decreased) concentrations relative to the cutoff level, the accuracy of test results depends on both the proximity of drug concentration to the cutoff level and analytical precision/accuracy. Therefore, the application of cutoff values to the interpretation of test results requires routine monitoring of analytical precision and accuracy.

Jain et al. (34) have addressed the fact that drug tests may be falsely reported as negative owing to statistical variation about the cutoff level by using reporting levels. The reporting level is the concentration for which the procedure can effectively give 100% accurate results under normal working conditions. For instance, if the analytic imprecision is a coefficient of variation of 20% at the cutoff concentration of 0.5 μg/mL, then the reporting level is selected as 0.8 μg/mL (0.5 μg/mL + 3 SD). By this design, Jain states that "the cut-off value then functions as a buffer to experimental error and/or analytical variation and insures that a drug at the reporting level is correctly identified in every case."

2.6. Quality Control

A quality-control program is implemented to monitor the performance of all analytical systems in use. The quality-control data generated are referenced to predefined limits of acceptability and when these limits are exceeded, appropriate measures are taken to remedy the deficiency.

There are three approaches to quality control that the laboratory should use to monitor and gauge its testing reliability: (1) "open" internal quality control using specimens with assay characteristics that are known to the analyst; (2) internal quality control using specimens that are "blind" to the analyst; and (3) external proficiency testing. If the quality-control material is to assess the performance of a method reliably, the behavior of quality-control specimens should mimic that of test specimens in the analytical procedure (35); drugs and analytically significant metabolites should be present in a matrix that is as nearly identical as possible to the matrix of the specimen being analyzed. This is especially important if the identity of the quality-control specimen is to be protected in a "blind" testing program.

Open Internal Quality-control Testing. This type of testing is implemented to provide a timely assessment of assay performance; the acceptability of an analytical run is determined by the analyst before test results are released for supervisory review. Suitable control materials must be available: (1) to verify that drugs/metabolites present in a specimen at or near a selected cutoff or reporting level are detected; (2) to verify that a true negative sample will demonstrate the absence of drugs; and (3) to monitor the accuracy, precision, and linearity of quantitative assays. Therefore, a minimum of three types of control materials should be available to the analyst: (1) a "threshold control" containing drugs/metabolites at or near the assay cutoff or reporting level; (2) a drug-free control (blank control); and (3) an "elevated control" containing drugs/metabolite at a concentration near the working, upper limit of assay linearity.

Each analytical system in use has its own requirements for quality-control testing. Guidelines specifying the nature of quality-control specimens, their frequency of assay, and their location among unknown test specimens within an analytical run should be established. These guidelines should provide a cost-effective means to monitor assay performance characteristics with a high detection capability for analytical error.

When samples are batched for analysis, unknown test samples are interspersed with control samples to measure analytical drift; controls of markedly different concentration, such as the "elevated" and blank controls, are analyzed sequentially in a run to measure carry-over or sample interaction. If sample interaction is an expected and documented phenomenon, as with gas chromatography column contamination by specimen extracts containing high concentration of drug, appropriate procedures are instituted to eliminate carry-over. The effectiveness of these procedures is then documented through analysis of a blank control.

Analytical drift is most effectively monitored at the assay reporting level. As described in Section 2.5, the reporting level for a drug is selected on the basis of known analytic variation; when the system is operating within acceptable limits of accuracy and precision, the reporting level control should always test positive. Should substandard analytical performance occur, the information available from the analysis of three levels of control (blank, threshold, and elevated) will provide an assessment of error type; that is, random versus systematic, constant versus proportional. When the analysis is deemed out of control, the run is rejected and the analyst must then document the source of error and the corrective action taken.

"Blind" Internal Quality-control Testing. This kind of testing is conducted to monitor each aspect of the testing process: chain of custody, accessioning, analysis, and reporting. To prevent a conflict of interest, the individual responsibile for this program should have no direct involvement with any phase of the testing process and should report only to laboratory administration. Clearly, blind quality-control testing, if conducted properly, is the most objective approach to obtaining unbiased information regarding the reliability of routine laboratory performance.

Blind control specimens differ from "open" controls in that they contain a drug or combination of drugs/metabolites that is representative of drug profiles commonly found in drug-positive client specimens; a blind control containing each drug in a test panel would readily be recognized as a check sample. A robust program includes over-the-counter and prescription medications that pose a challenge to the analyst for accurate differentiation from the drug of interest. Because sample handling and reporting procedures are being monitored in addition to analytical performance, the sample must be introduced into the laboratory as a client specimen. The logistics of the procedure may require the program coordinator to provide a client with controls for batching with unknown test specimens; this arrangement should be acceptable if proper credits are applied to the client's account. The frequency of this mode of quality-control testing is largely dictated by the resources available to the quality-control coordinator; however, blind testing should be conducted at least once each day of laboratory operation. This would permit a timely investigation of testing errors and their correction before reports are released to the client.

2.7.   Reporting

At the completion of the analytical process, it is necessary to collate all records of the analysis to verify completeness and accuracy. In effect, the reviewer (certifying scientist) duplicates the decision process previously performed by the bench analysts and certifies that the analytical data support the stated test findings. The review process must be fully documented, and this is most effectively achieved by applying a checklist to data generated from specimen analysis. The checklist should outline the items to be reviewed and the respective standards of acceptability. Should there be a violation of a standard, the appropriate staff member must be notified and an acceptable course of action must be documented before the test result is released.

The test report should clearly document the drug(s) and/or metabolite(s) tested for, the concentration below which test results are reported as negative (threshold concentration), the test result, and if the test finding is positive, the method used for confirmation. Because the drugs are listed, a negative report is a specific finding and does not mislead the physician as to the scope of testing performed. Because laboratories often support more than one method that can be used for confirmation, it is appropriate that the method used be specified on the report form. Confidentiality of test reports must be ensured. Reports of test findings should be made available only to the individual authorized to request the analysis. Access to laboratory facilities where records are stored must be restricted to authorized personnel.

3.   ANALYTICAL STANDARDS AND CONTROL MATERIALS

The inventory of pure drugs commercially available to the toxicologist for standard and control preparation has increased significantly in the recent past. Increased demand for toxicology services by drug treatment and testing programs will likely produce an even greater availability of authentic drugs, metabolites, and isotopically labeled standards. The analytical laboratory, however, must hold appropriate state and federal licenses to purchase scheduled substances. Small quantities of drug standard in solvent (generally 1 mg/mL) are available from most chromatography supply houses to laboratories without the licenses.

Information on drug purity, solubility, chemical characteristics, and storage conditions is often provided by the distributor. If the information is lacking or additional data are required, the distributor will provide file data for the lot of material in question. Nonetheless, the laboratory should generate an internal record of the physical–chemical properties (e.g., melting or boiling point, ultraviolet spectrum, mass spectrum) of the drug to confirm the state of purity prior to use as an analytical standard. It is not uncommon for a laboratory to store drugs for many months or years; a periodic reanalysis of physical—chemical properties can be referenced to original records as a check on drug stability. Likewise, the ultraviolet spectra of freshly prepared and ''aging'' stock solutions of drug provide useful information on the accuracy of preparation and changes in concentration due to solvent evaporation or degradation of drug.

The preparation of analytical standards and calibration curves is an exacting process. A known amount of analyte is added to a known volume of matrix that has been certified to be free of both analyte and substances that interfere with the assay. The matrix used in preparation should be as nearly identical as possible to that of the test specimen; the medium used to supplement matrix with analyte and the agents used to ensure the stability of analyte and matrix should be evaluated for possible matrix related effects that may adversely affect assay performance (see related discussion on quality-control materials, Section 2.4).

Before an analytical method is adopted by the laboratory for routine use, it must undergo an extensive performance evaluation. Analytical performance characteristics that are important to the analyst include accuracy, recovery, sensitivity, specificity, precision, linearity, ease of use, and cost (17, 18). Once determined, the performance characteristics are compared to predetermined criteria of acceptability and an informed decision of compatibility with laboratory expectations and needs is made (36).

The task of determining assay accuracy is a difficult one, primarily because Standard Reference Materials (SRM, National Bureau of Standards) and primary standards are not generally available to the analytical toxicology laboratory and definitive methods have yet to be developed. An alternative approach, although not as robust, is the use of survey-validated reference material (SVRM) (37). Proficiency testing programs, such as that sponsored by the College of American Pathologists, produce survey materials that are assayed by many hundreds of laboratories. The consensus mean concentration of test results indicates the current state of the art in determining the true amount of an analyte in a biological matrix. When consensus means are compared to results produced by reference methods, there is generally excellent agreement and the materials can be viewed as secondary standards (38, 39).

Application of linear regression analysis to analytical data produced from the analysis of SVRMs therefore provides useful information on the accuracy of method standardization. As an example, survey materials produced by the New York State Department of Health drug monitoring/ quantitative toxicology program were assayed for phenytoin by an HPLC method developed in my laboratory. The consensus mean of proficiency survey test results, verified by reference laboratory analyses, was used for target assignment for each of the 18 specimens included in the regression analysis. The results of the analysis are provided in Fig. 1.1. The 95% confidence intervals about the slope and intercept include the ideal values of 1.0 and 0.0, respectively; hence the conclusion that proportional and constant error does not exist might be drawn. However, the slope and intercept are highly correlated and it is necessary to evaluate the joint confidence intervals (40, 41). The joint confidence intervals of slope and intercept are displayed as an ellipse (see Fig. 1.2), and permissible values of each are included within this ellipse. As can be seen, the intersection of slope = 1.0 and intercept = 0.0 falls within the ellipse and it can be safely concluded that the analytical method is free of both proportional *and* constant systematic errors.

4. QUALITY-CONTROL MATERIALS

Once an analytical method has been adopted for routine use, a quality-control program is implemented to monitor long-term assay precision and accuracy (see Sections 1.6 and 2.6). The success of the program depends on the selection and appropriate use of control materials, timely and thorough documentation of control data, criteria used for interpretation of control data, and the commitment and expertise of the analyst to identify and remedy substandard assay performance.

For reasons previously described (see Sections 1.6 and 2.6), drug-free (negative), threshold, and "elevated" control materials should be available to the analyst. Because the toxicology laboratory generally supports several analytical techniques, each with its own limit of detection and/or cutoff concentration, the task of procuring appropriately designed quality-control materials is a difficult one. Commercial availability of control materials is less than adequate; hence the toxicologist must often prepare controls in-house.

The desired characteristics of control material include (1) matrix iden-

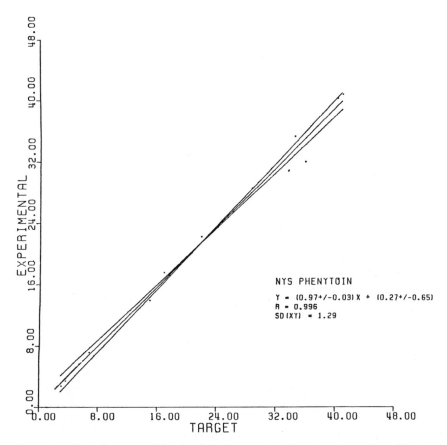

Figure 1.1. Use of survey-validated reference materials (SVRM) to determine calibration accuracy of an HPLC serum phenytoin assay. Linear regression analysis of HPLC phenytoin data (experimental) was performed on the "true" phenytoin concentration (target) determined as the consensus mean of New York State (NYS) program participant data. R is the regression coefficient; SD (xy) is the standard error estimate.

tical, or as nearly as possible, to that of the test specimen; (2) known concentration of analyte; (3) analyte concentration at critical point(s) within the analytical range, for example, the limit of detection, cutoff, or reporting level, as appropriate; (4) stability of both the analyte and the matrix; and (5) ample supply. Matrix effects on the performance of an assay are often unpredictable and difficult to control. The control matrix therefore should have characteristics very similar to that of the test specimens and should be modified during control preparation only as needed to ensure stability (35).

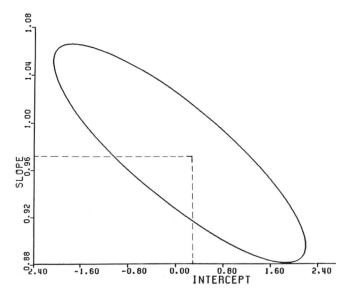

Figure 1.2. Joint 95% confidence interval for least squares estimates of slope (m) and intercept (b) obtained from regression analysis of phenytoin data (see Fig. 1.1).

The need for different levels of analyte concentration can be addressed by serial dilution of an "elevated" control with drug-free matrix. A laboratory may use any one or a combination of TLC, EIA, and GC/MS methods for analysis. Appropriately designed control materials provide the means to (1) monitor analytical response to a drug-free specimen, (2) monitor accuracy and precision near the cutoff or reporting level of EIA and GC/MS methods, (3) validate the TLC detectability of drug at the limit of detection, and (4) monitor the effectiveness of procedures to eliminate carry-over by placing a negative control immediately after the "elevated" control in the analytical run. The serially diluted specimens can be used routinely to monitor assay performance throughout the analytical range of quantitative and semiquantitative procedures. If performance should deviate from an acceptable level, quality-control data will provide a means to determine the nature of error, that is, systematic (constant or proportional) or random.

5. DOCUMENTATION OF CONTROL DATA

The quality-control program is a major expense to the laboratory. The purchase of commercial controls or materials to prepare controls, participation in proficiency testing or external quality-control programs, reagents, time to process control specimens, and time to document and assimilate quality-control data all factor into the determination of laboratory expense. If documentation of data is not complete or if presentation of data is not amenable to efficient and reliable interpretation, laboratory resources are wasted and the quality of the testing service suffers.

Qualitative control data such as that generated by thin-layer chromatography and spot tests provides an immediate assessment of the acceptability of an analytical run. Although the data are not subjected to statistical methods of analysis or to interpretive criteria other than "detected" or "not detected," the need to document the control test results must be impressed upon the analyst. If control test results are not documented, there is a justified assumption that the control specimens were not processed. Relevant parameters to record include the date of assay, the identification of control specimens and test specimen(s) included in the analytical run, the expected and observed control test findings, a notation of conformance with expected assay characteristics (e.g., color, intensity, R_f), and identification of the analyst.

Quantitative methods of analysis require statistical quality control. As described by Westgard and Klee (42), the best control procedure is one with the lowest probability for false rejection of an analytical run and the highest probability for detecting those errors that are large enough to invalidate the analytical quality goals for the method. Levey–Jennings (43), cumulative sum (44), Shewhart mean and range (45), and Westgard multi-rule (46) control procedures represent the most common approaches to statistical quality control in the clinical chemistry laboratory. The procedures vary in complexity and the laboratory should take into consideration the skills of the analysts and availability of computer resources before a control procedure is selected.

In our laboratory, a computer is used to document control data (see Fig. 1.3) and to generate Levey–Jennings charts for inspection of the data (see Fig. 1.4). The Westgard "multi-rule" control rules are used for data interpretation (see Fig. 1.5). As control data are entered, analytical variables that provide additional insight into assay performance are also documented. In the case of liquid and gas chromatography procedures, the column efficiency and peak symmetry are determined and recorded. Should an "out-of-control" situation arise, the control rule that was violated and records of both the control data and column performance characteristics can be used to identify more readily the source of error and suggest a course of corrective action.

Analyte: Acetaminophen

Lot ID	Target	Std. Dev.
C4	197.4	5.94
C5	100.8	4.02
C7	27.5	1.51

assay #	Date	C4	C5	C7	Eff	Sym	Violation	Comment
1	02/14/86	200.0	102.0	27.5	2930	1.0	No	Altex Col 2UE5851N
2	03/24/86	196.3	103.4	28.6	5324	1.1	No	Repaired Column
3	04/15/86	198.2	101.0	25.9	6946	1.1	No	
4	05/22/86	202.8	108.5	27.3	3579	1.2	No	
5	06/27/86	199.8	104.5	26.2	2967	1.3	No	Plumbing change
6	07/21/86	192.6	94.0	24.5	5673	1.1	No	Col change 4UE4716N
7	08/25/86	210.0	104.0	28.4			No	
8	09/12/86	195.0	98.3	27.5			No	
9	10/29/86	199.6	100.6	27.5	2984	1.1	No	
10	12/03/86	196.7	100.0	27.9	2125	1.2	No	
11	12/31/86	195.2	101.2	27.6	6143	1.0	No	Repaired Col & seals
12	01/22/86	192.9	99.1	27.0	5960	1.0	No	
13	03/04/87	186.4	94.0	26.9	2568	1.2	No	
14	04/01/87	191.2	98.9	26.5	10270	1.2	No	Perkin-Elmer column
15	05/06/87	196.6	98.2	28.4	4190	1.1	No	
16	07/24/87	197.5	100.0	29.9	5852	1.0	No	
17	10/22/87	184.6	102.0	22.8	3842	1.1	R:4s	poor col eff
18	10/23/87	201.2	102.2	29.0	10074	1.1	No	New col cartridge
19	01/12/88	188.7	96.7	26.2	10755	1.1	No	
20	01/13/88	199.8	101.0	29.6			No	

....

Location and dispersion estimates for assays 1 through 20

	C4	C5	C7
Location:	196.4	100.5	27.4
Dispersion:	5.48	3.16	1.46
CV (%):	2.8	3.1	5.3
Number:	20	20	20

Figure 1.3. Computer-generated documentation of internal quality-control data for high performance liquid chromatographic determinations of serum acetaminophen concentration. Eff is column efficiency; sym is peak symmetry; violation refers to Shewhart "multi-rule" control analysis (see Fig. 1.5).

6. STANDARD OPERATING PROCEDURE MANUAL

The contribution to preanalytic and analytic error is likely to be significant if standard procedures for specimen collection and handling and execution of analytical methods are not followed. It is necessary, therefore, to clearly state in a standard operating procedure manual (SOPM) accepted procedures for each component of the test process. The information provided in the SOPM must be accurate and sufficiently complete to allow independent and reliable analyses.

The NCCLS has provided a standard for the organization and content of the SOPM or technical procedure manual (47) (see Table 1.1). The

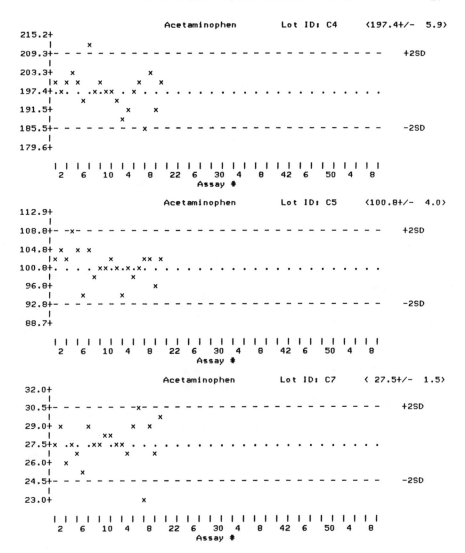

Figure 1.4. Levey–Jennings presentation of acetaminophen control data listed in Fig. 1.3.

information that the NCCLS specified to be included in the SOPM is basic to the quality assurance of testing service and each item should be fully addressed. It is essential that each procedure in the manual be kept up to date and that the director or qualified staff review the manual at least annually to ensure that recorded procedures are consistent with laboratory practice. Such reviews and updates should be documented by au-

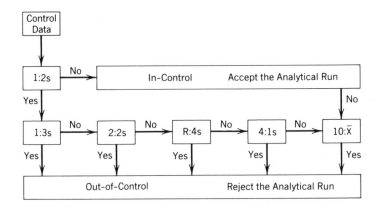

Violation	Error Type	Potential Causes	Corrective Measures
1:3s	Random; systematic	Many possibilities	Repeat the entire analytical procedure. If the problem persists, review all possibilities listed below
2:2s	Random; systematic	(1) Poor chromatography, (2) leaking piston seals, (3) variable injection volume, (4) detector instability, (5) calibration, (6) control stability	(1) Check column efficiency; (2) check piston salt deposition; (3) check loop patency; (4) check noise level, change desiccant; (5) prepare fresh extraction solvent, evaluate calibrators; (6) prepare fresh controls
4:1s	Systematic	(1) Calibration, (2) extraction	(1) Check calibration with SVRM, prepare new standards if necessary; (2) prepare fresh extraction solvent
R:4s	Random	Review 2:2s	Review 2:2s
10:$\overline{X}$	Systematic	Review 4:1s	Review 4:1s

Figure 1.5. Quality Control Rules for Deciding the Acceptability of an Analytical Run. Multi-Rule Shewhart Analysis.

Table 1.1. Information Content and Organization of the Laboratory Standard Operating Procedure Manual (47).

Procedure Title

Substance tested for and the specimen matrix
Method of analysis

Principle

Description of analytical principle
Application for testing

Specimen Required

Required patient preparation
Preferred type of specimen
Amount of specimen required (maximum and mimimum)
Collection containers, preservatives, anticoagulant
Specimen stability and storage requirements
Criteria for unacceptable specimens and course of action
Specimen characteristics that might compromise the analysis (e.g., hemolysis, lipemia, icteric coloration)

Reagents—Special Supplies and Equipment

Equipment and supplies used
Supply company for reagent kits and prepared reagents
Directions for reagent preparation
Procedures used to determine acceptable reagent performance
Storage conditions
    Container
    Temperature
    Stability
    Labeling (substance, lot number, preparation and expiration dates, prepared by)

Calibration

Standard preparation
Calibration procedure

Quality Control

Control materials used
Instructions for preparation and handling
Frequency of assay
Description of how tolerance limits are established
Corrective action when tolerances are exceeded
Description of how data is to be recorded and stored
Alternatives if controls are not available

Table 1.1. (*Continued*)

Procedure

Stepwise instruction on use of materials, reagents, supplies, and instrumentation in performing an analysis

Calculations

Stepwise instruction with examples

Reporting Results

Reference range(s), if applicable
See Sections 1.6 and 2.7

Procedural Notes

Additional information that will provide insight to requirements for reliable analyses (e.g., sources of error, reasons for special precautions)

Limitations of Procedure

Linearity and/or detection limits
Known interferring substances

References

Sources of information used to establish standard operating procedure

Supplemental Materials

Product literature, flow diagrams, etc. that might be used at the bench

---

thorized signature and date. Although not generally done, it is also appropriate to have each analyst document—by signature and date—that the procedures and updates have been reviewed and understood.

7. LABORATORY LICENSURE AND ACCREDITATION

Laboratory licensure and accreditation programs are an important component of the laboratory's aggregate quality assurance program. Participation in these programs provides an objective means to evaluate the quality of testing service. Laboratories are held to administrative and analytical standards, designed to ensure a quality of testing that is consistent with medical requirements for good patient care. Whether participation in the external evaluation program is mandated or voluntary, the proper perspective should be that the agency or professional organization is providing a service to the laboratory and to consumers of laboratory testing services. Specifically, program objectives are to document and improve the quality of routine laboratory practice; if deficiencies are identified, the conscientious laboratory will take corrective action and therefore benefit from the educational aspect. In those instances where the quality of testing is substandard and the laboratory has not responded favorably to remediation efforts, patient care may be jeopardized and continued testing must be restricted.

The objectivity of the licensure/accreditation program is sustained only in those instances where (i) the participant allows review of laboratory practice as it is routinely conducted and (ii) proficiency testing specimens are treated as patient specimens. Unfortunately, there are numerous accounts of lesser quality of testing when proficiency specimens are submitted "blind" to the analyst (48–52). In those instances where the laboratory "prepares" for an inspection or processes test specimens with undue consideration, the licensure/accreditation program documents the highest quality of testing environment and performance that the laboratory is capable of providing. The laboratory director is deprived of useful input as it pertains to routine laboratory practice and the licensure/accreditation program may incorrectly assess laboratory qualifications.

At present, relatively few proficiency testing and inspection programs are available to the toxicology laboratory. The College of American Pathologists (CAP) supports the only national accreditation program in clinical toxicology: both the CAP and the National Institute on Drug Abuse have recently made available accreditation programs in forensic urine drug testing. State Health Departments in New York and Pennsylvania maintain a laboratory inspection and proficiency testing program and require all laboratories doing business in the state—whether located in or out of state—to participate in the licensure program. Many other states require toxicology laboratories in their jurisdiction that perform workplace drug testing to participate in an approved proficiency testing program. Licensure requirements vary significantly among states and the laboratory must obtain the appropriate permits before testing is performed on specimens obtained from residents within the jurisdiction.

Proficiency testing programs are also available from the American Association for Clinical Chemistry and the American Association of Bioanalysts. The Center for Disease Control (CDC) has recently issued a proposal for a Uniform Proficiency Testing Program for Medicare-certified laboratories and laboratories licensed under the Clinical Laboratories Improvement Act of 1967 (53). The CDC objective is to define a proficiency testing program that could be used as the national standard for evaluating the performance of laboratories. Unfortunately, the discipline of analytical toxicology was not addressed in this proposal. Proficiency testing programs therefore will likely continue to vary considerably in their degree of challenge and applicability to types of testing service, that is, clinical toxicology versus substance abuse testing versus forensic toxicology. The toxicology laboratory should participate in as many external quality-control programs as are relevant to the type of service it offers.

REFERENCES

1. C. B. Walberg, V. A. Pantlik, and G. D. Lundberg, *Clin. Chem., 24,* 507 (1978).

2. R. L. Taylor, S. L. Cohan, and J. D. White, *Am. J. Emerg. Med., 3,* 504 (1985).

3. M. A. McGuigan, *Clin. Symp., 36,* 3 (1984).

4. W. L. Thompson, "Recognition, treatment and prevention of poisoning," in W. C. Shoemaker, W. L. Thompson, and P. R. Holbrook, Eds., *Textbook of Critical Care,* Saunders, Philadelphia, 1984, p. 801.

5. B. R. Hepler, C. A. Sutheimer, and I. Sunshine, *Med. Toxicol., 1,* 61 (1986).

6. R. P. Bateh, *Clin. Lab. Med., 7,* 371 (1987).

7. NCCLS Tentative Standard TST/DM-1, *Standard for the Development of Requisition Forms for Therapeutic Drug Monitoring and/or Overdose Toxicology,* National Committee for Clinical Laboratory Standards, Villanova, PA, 1980.

8. A. R. Henderson, *J. Clin. Pathol., 35,* 986 (1982).

9. B. R. Hepler, C. A. Sutheimer, and I. Sunshine, *J. Toxicol. Clin. Toxicol., 19,* 353 (1982).

10. R. T. Chamberlain, *Clin. Biochem., 19,* 122 (1986).

11. C. R. Hamlin, *Clin. Chem., 34,* 158 (1988).

12. Clinical Laboratory Handbook for Patient Preparation & Specimen Handling, Fascicle IV, *Therapeutic Drug Monitoring and Toxicology,* College of American Pathologists, Skokie, IL, 1985.

13. M. A. Pfaller, G. G. Granich, R. Valdes, and P. R. Murray, *Diagn. Microbiol. Infect. Dis., 2,* 93 (1984).

14. M. I. Walters and W. H. Roberts, *Ther. Drug Monit., 6,* 199 (1984).

15. C. A. Queen and C. S. Frings, *Clin. Chim. Acta, 45,* 307 (1973).

16. National Institute on Drug Abuse, Statistical Series, Annual Report, *Data from the Drug Abuse Warning Network (DAWN),* U.S. Department of Health and Human Services, Rockville, MD, 1985.

17. L. G. Nielsen and K. O. Ash, *Am. J. Med. Technol., 44,* 30 (1978).

18. J. O. Westgard, D. J. de Vos, M. R. Hunt, E. F. Quam, R. N. Carey, and C. C. Garber, *Am. J. Med. Technol., 44,* 420 (1978).

19. R. W. Jenny, *Clin. Chem., 31,* 1158 (1985).

20. B. H. Rumack and H. Matthew, *Pediatrics, 55,* 871 (1975).

21. F. Wilson, *Lab. Management, 22,* 45 (1984).

22. W. Pollin, *Issues Sci. Technol., 3,* 20 (1987).

23. J. E. Manno, "Specimen Collection and Handling," in R. L. Hawks and C. N. Chiang, Eds., *Urine Testing for Drugs of Abuse,* NIDA Research Monograph 73, National Institute on Drug Abuse, Rockville, MD, 1986.

24. C. S. Frings and C. A. Queen, *Clin. Chem., 18,* 1442 (1972).

25. M. K. Googins and F. S. Apple, *Clin. Chem., 33,* 972 (1987).

26. Y. Liu, R. D. Budd, and E. C. Griesemer, *J. Chromatogr., 248,* 318 (1982).

27. R. C. Baselt, *J. Chromatogr., 268,* 502 (1983).

28. S. Kaye, *Am. J. Clin. Pathol., 74,* 743 (1980).

29. Department of Health and Human Services, *Fed. Regist., 52,* 30638 (1987).

30. D. W. Hoyt, R. E. Finnigan, T. Nee, T. F. Shults, and T. J. Butler, *J Am. Med. Assoc., 258,* 504 (1987).

31. U. Boerner, S. Abbott, and R. L. Roe, *Drug Metab. Rev., 4,* 39 (1975).

32. R. L. Hawks and C. N. Chiang, "Examples of Specific Drug Assays," in R. L. Hawks and C. N. Chiang, Eds., *Urine Testing for Drugs of Abuse,* NIDA Research Monograph 73, National Institute on Drug Abuse, Rockville, MD, 1986, p. 98.

33. A. D. Fraser, *J. Anal. Toxicol., 11,* 263 (1987).

34. N. C. Jain, T. C. Sneath, and R. D. Budd, *J. Anal. Toxicol., 1,* 142 (1977).

35. R. Rej, R. W. Jenny, and J-P. Bretaudiere, *Talanta, 31,* 851 (1984).

36. J. O. Westgard, D. J. de Vos, M. R. Hunt, E. F. Quam, R. N. Carey, and C. C. Garber, *Am. J. Med. Technol., 44,* 803 (1978).

37. G. F. Grannis, *Pathologist, 32,* 96 (1978).

38. G. F. Grannis, *Clin. Chem., 22,* 1027 (1976).

39. P. M. G. Boughton and L. Eldjarn, *Ann. Clin. Biochem., 22,* 625 (1985).

40. J. Mandel and F. J. Linnig, *Anal. Chem., 29,* 743 (1957).

41. R. B. Davis, J. E. Thompson, and H. L. Pardue, *Clin. Chem., 24,* 611 (1978).

42. J. O. Westgard and G. G. Klee, "Quality Assurance," in N. W. Tietz, Ed., *Textbook of Clinical Chemistry,* Saunders, Philadelphia, 1986, p. 424.

43. S. Levey and E. R. Jennings, *Am. J. Clin. Pathol., 20,* 1059 (1950).

44. J. O. Westgard, T. Groth, T. Aronsson, H. Falk, and C-H. de Verdier, *Clin. Chem., 23,* 1857 (1977).

45. A. Hainline, "Quality Assurance: Theoretical and Practical Aspects," in W. R. Faulkner, Ed., *Selected Methods in Clinical Chemistry,* Vol. 9, American Association for Clinical Chemistry, Washington, DC, 1982, p. 17.

46. J. O. Westgard, P. L. Barry, M. R. Hunt, and T. Groth, *Clin. Chem., 27,* 493 (1981).

47. NCCLS Approved Guideline, GP2-A, *Clinical Laboratory Procedure Manuals,* National Committee for Clinical Laboratory Standards, Villanova, PA, 1984.

48. E. Gottheil, G. R. Caddy, and D. L. Austin, *J. Am. Med. Assoc., 236,* 1035 (1976).

49. W. McCormick, J. A. Ingelfinger, G. Isakson, and P. Goldman, *N. Engl. J. Med., 299,* 1118 (1978).

50. D. J. Boone, H. J. Hansen, T. L. Hearn, D. S. Lewis, and D. Dudley, *Am. J. Public Health, 72,* 1364 (1982).

51. J. A. Ingelfinger, G. Isakson, D. Shine, C. E. Costello, and P. Goldman, *Clin. Pharmacol. Ther., 29,* 570 (1983).

52. H. J. Hansen, S. P. Caudill, and D. J. Boone, *J. Am. Med. Assoc., 253,* 2382 (1985).

53. Division of Assessment and Management Consultation, *CDC Proposal for a Uniform Proficiency Testing Program,* U.S. Department of Health and Human Services, Center for Disease Control, Atlanta, GA, 1987.

CHAPTER

2

EMIT ASSAYS FOR DRUGS OF ABUSE

BARBARA A. SMITH AND JEAN C. JOSEPH

St. Mary Medical Center
Long Beach, California

1. GENERAL

1.1. Principle of EMIT® Assays

Immunoassay techniques utilize antibodies specific for particular proteins or other large and small molecules (antigens). A competition occurs between the antigen to be quantitated and the antigen attached to a chemical (label) entity (label) that enables it to be detected by the measurement of radioactivity, fluorescence, or enzyme activity. The greater the quantity of antigen present in the sample, the less labeled antigen will combine with the antibody.

35

In the 1960s, radioimmunoassay (RIA) was at the forefront of the measurement of proteins. This methodology is defined by the use of radioactive material as the label and necessitates separating bound from free label before measurement of radioactivity can be accomplished. The RIA technique is now used routinely in the military for drug-of-abuse screening.

In 1972, EMIT® assays (enzyme multiplied immunoassay technique) were introduced to the laboratory market by Syva Company. Drugs of abuse were the first analytes to be measured. The opiate and methadone assays were made available for screening in methadone treatment clinics, clinical and commercial laboratories, and military institutions. The first EMIT assays were commercially available to perform therapeutic drug monitoring. Assays of antiepileptic drugs—phenobarbital, phenytoin, primidone, carbamazepine, and ethosuximide—were introduced (1).

The advantages of enzyme immunoassay include no use of radioactive material and thus no radioactive licensing or radioactive disposal, as well as a longer shelf life for reagents. The use of spectrophotometric measurements makes assays adaptable to numerous instruments on the market today. The EMIT assay requires no separation of bound from free fractions. Urine or serum is mixed with the specific antibody and the enzyme substrate. A drug labeled with an enzyme is then added. Competition occurs between the drug present in the sample and the enzyme-labeled drug for the antibody binding sites. When the enzyme-labeled drug binds to the antibody, the enzyme is conformationally blocked. The enzyme activity of the reaction mixture subsequently measured is therefore the enzyme activity derived from the free enzyme-labeled drug alone.

Another commercial assay system which does not require separation steps or radioactivity is the fluorescence polarization immunoassay developed by Abbott Laboratories. Assays are now available for most of the drugs of abuse using the Abbott TDx or ADx instrument.

The lysozyme enzyme system (2) was the first enzyme system introduced for the EMIT assay. Untreated urine plus the mucopolysaccharide portion of the cell wall of the bacteria M. leuteus, acting as the enzyme substrate, and Reagent A, containing antibodies to the compound to be tested, were mixed (3). Reagent B, containing a drug labeled with the enzyme lysozyme, was then added. The enzyme activity was measured spectrophotometrically at 436 nm. The measured enzyme activity was read initially at 10 s and again at 50 s after it was placed in a tightly defined, thermostatically controlled flow cell at $37 \pm 0.1°C$. As the bacteria M. leuteus was lysed, the absorbance reading decreased. The rate of decreased absorbance measured was proportional to the concentration of drug and/or metabolite present in the sample.

The disadvantages of the lysozyme assays were related to the numerous variables in each assay. As the bacterial suspension *M. leuteus* acting as the substrate aged, the rate differences between the negative and low calibrators decreased, thereby decreasing sensitivity. A potential source of false positives was endogenous lysozyme excreted in the urine by 2–4% of the population. It was necessary that a urine blank be performed on all positive samples to uncover the presence of endogenous lysozyme. If endogenous lysozyme was found to be present, it was necessary to find an alternative methodology. These disadvantages are not seen with the glucose-6-phosphate dehydrogenase (G6PDH) enzyme system, which has replaced the lysozyme enzyme system.

In the G6PDH assay, Reagent A contains antibody to the drug, the substrate glucose-6-phosphate, and the coenzyme NAD. Reagent B contains a drug labeled with the enzyme G6PDH. Reagent A is added first, and the drug in the patient sample binds to the antibody. Reagent B containing the enzyme-labeled drug is added next, and the enzyme-labeled drug binds to any remaining antibody sites, thus blocking the enzyme's activity. Any enzyme remaining unbound catalyzes the reaction of G-6-P with NAD to form NADH, which is measured spectrophotometrically at 340 nm. The measured enzyme activity is read initially at 15 s and again at 45 s after it is placed in a tightly defined, thermostatically controlled flow cell at 30 ± 0.1°C. The increase in absorbance at 340 nm is proportional to the concentration of drug and/or metabolite present in the sample.

At present, amphetamines, barbiturates, benzodiazepines, cannabinoids, cocaine metabolite, methadone, methaqualone, opiates, propoxyphene, and phencyclidine are available as EMIT-d.a.u.™ assays and EMIT-st™ urine assays. Benzodiazepines, barbiturates, and tricyclic antidepressants are available as EMIT-tox serum and EMIT-st serum assays. Serum acetaminophen is available as an EMIT-tox serum assay and ethyl alcohol as an EMIT-st serum assay.

1.2. Specimen Collection and Storage

Because of the legal implications of the results of drug of abuse analysis, many means of sample adulteration are utilized by drug abusers in order to decrease the detection of drugs in the sample (4). One means used is dilution of the specimen. It is important that the urine specific gravity be measured, because the sample may have been diluted by the use of diuretics or the addition of water. The specific gravity coupled with the sample temperature will help detect a diluted specimen.

The pH of the urine should be measured. The enzyme immunoassay is designed to perform in a pH range of 5.0–8.0 (5), or in the case of the cannabinoid assay, a pH of 5.5–8.0 (6). The 0.055 Tris-HCl buffer can adequately handle normal urine samples. The pH of urine samples can be adjusted to neutral by the addition of 1 mol/L HCl or 1 mol/L NaOH.

Specimens displaying unusually high turbidity should be centrifuged prior to analysis in order to decrease the initial absorbance reading (A_0). It is unknown at this time whether the stability and integrity of the drugs are affected by bacterial contamination. The effects of preservatives are also unknown and are therefore not recommended.

It is recommended that freshly voided urine, for assays other than cannabinoids, be stored in glass or plastic containers and be assayed within 3 days when stored at 2–8°C. Storage of specimens with drug concentrations at or near the low calibrator may exhibit negative results with prolonged storage. Drugfree specimens have not assayed as positive after prolonged storage.

Specimens to be analyzed for cannabinoids, if not analyzed within 24 hr, should be stored frozen at approximately −12°C (6). The specimen should be thawed and brought to room temperature (20–25°C) before analysis. Prolonged storage in plastic containers, direct sunlight, or elevated temperatures may cause deterioration of the sample.

1.3.   Assay Conditions

Properly functioning instrumentation is very important in performing reliable EMIT assays. The buildup of protein material on tubing can denature enzymes in some instances. Accurate sample pipetting devices are necessary to ensure proper proportions and mixing of reagents, buffer, and sample.

Assay reagents should be at room temperature (20–25°C) for testing (5). The reconstituted reagents are stable for 12 weeks when stored at 2–8°C.

The qualitative/semiquantitative drug of abuse and serum toxicology assays require the use of external or in-house quality-control material. The control should contain an adequate concentration of each drug to be assayed in order to produce a positive response.

The initial absorbance reading should be monitored because it ensures proper instrument settings and the integrity of the assay on the instrument used. Once a day, the negative, low, and medium calibrators should be run. The differences between the changes in absorbances of the negative and low ($\Delta A_{low} - \Delta A_{neg}$) and the low and medium ($\Delta A_{med} - \Delta A_{low}$) calibrators ensure the performance of each assay, and the minimum differences are listed in each package insert (5–17). For qualitative testing,

Table 2.1. **Quality-Control Parameters for Calibrators**

Assay	$\Delta A_{low} - \Delta A_{neg}$	$\Delta A_{med} - \Delta A_{low}$	Min A_0	Max A_0
Amphetamine	55	60	800	2000
Barbiturate	45	45	600	1800
Barbiturate (serum)	20	40	300	1800
Benzodiazepine	15	35	600	1800
Benzodiazepine (serum)	20	140	300	1800
Cannabinoid 100 ng	20	35	800	1800
Cocaine metabolite	25	65	900	1500
Methadone	20	40	600	1800
Methaqualone	35	55	400	1500
Opiate	20	30	600	1800
Phencyclidine	20	40	500	2000
Propoxyphene	15	50	800	1800
Tricyclic antidepressants (serum)	25	50	500	1500

the low calibrator should be run in duplicate every 4 hr, with results agreeing within 6 absorbance units. Samples run in duplicate should also agree within 6 units. Table 2.1 lists quality-control parameters for the drug of abuse assays.

1.4. Instrumentation

Numerous instruments are available for use with EMIT assays. Besides the Syva AutoCarousel™ and Syva ETS™, drug of abuse assay applications are available for many different instruments, such as the ABA 100 manufactured by Abbott Laboratories, the Cobas-Bio by Roche Diagnostics, the Hitachi 705 by Boehringer Mannheim Diagnostics, the Optimate by Ames, the Multistat and Monarch by Instrumentation Laboratories, and many others. Information on applications of EMIT assays to these and other instruments may be obtained from Syva Company and the individual instrument manufacturers (1).

Syva manufactures an instrument, the Syva st™ System, for analyzing samples on a stat basis. A reagent system consisting of one test reagent vial is available. In the following sections on individual drug assays, these reagents may occasionally be mentioned when a particular interference is discussed; for the most part, however, the reagents being discussed are the EMIT d.a.u. formulations.

2. INDIVIDUAL DRUG ASSAYS

2.1. Amphetamines

Amphetamine has been used as a general term to describe synthetic ephedrine derivatives (18, 19). Structurally, the compounds contain a phenyl group with an amino group on the side chain. The addition of hydroxyl groups on the benzene ring or the side chain or substitutions on the nitrogen atom influence the drug's absorption, metabolic rate, and action on its receptors. Four classifications exist based on clinical usage: vasoconstrictors include phenylephrine, phenylpropanolamine (PPA), and mephentermine; vasodilators include nylidrin and isoxsuprine; bronchodilators include ephedrine, pseudoephedrine, and methoxyphenamine; and stimulants and anorexiants include amphetamine, methamphetamine, phentermine, and phenmetrazine.

All these drugs are available by prescription or over the counter. Drug abuse of these substances began in the 1940s and is still prevalent today. There is mild physical dependence with amphetamine abuse, but marked tolerance is seen, and high doses may result in toxic psychosis. The diagnosis of toxicity relies on observations of irritability, hyperactivity, diaphoresis, mydriasis, and hypertension. Street names are numerous and include "bennies," "speed," and "crystal meth." Amphetamine is well absorbed from the gastrointestinal tract. Metabolism occurs by deamination and hydroxylation. Deamination is the predominant route, with the main metabolite being phenylacetone. The accumulation of hydroxylated metabolites has been implicated in the development of psychosis. The oxidation of deaminated metabolites results in the formation of benzoic acid and hippuric acid. Amphetamine excretion is pH dependent. At a pH of less than 5.6, more than 50% of the dosage will be excreted, with a half-life of 7–8 hr. As the pH becomes more alkaline, as little as 3% will be excreted, with a half-life equal to 18–33 hr. Studies have shown detection 24–48 hr after a single ingestion.

The EMIT Amphetamine Assay (5) screens for the drug class sympathomimetic amines. The sheep antibody is most sensitive to amphetamine and methamphetamine. The low and medium calibrators contain 0.3 and 2.0 mg/L of d,l-amphetamine, respectively. The following drugs at the indicated concentrations cause positive results: methamphetamine (1.0 μg/L) and d-amphetamine (0.7 mg/L).

Phenethylamines, being structurally similar to amphetamine and methamphetamine, elicit a positive result with the amphetamine assay. The differentiation of these compounds can be accomplished by altering the chemical structure, especially those structures containing hydroxyl groups proximal to the amino group (i.e., ephedrine, PPA). Sodium periodate acts as an oxidant. Carbon–carbon bonds are cleaved at an aliphatic chain, or oxidative deamination occurs if the compound is susceptible to the basic solution. Amphetamine and methamphetamine are not affected. The EMIT Amphetamine Confirmation Kit (20) may distinguish PPA and ephedrine from amphetamine and methamphetamine. If the concentration of ephedrine or PPA is higher than the capacity of the sodium periodate to oxidize them, they may yield a positive amphetamine test. Isoxsuprine, mephentermine, nylidrin, phenmetrazine, phentermine, and diethylpropion cannot be distinguished from amphetamine by the Syva Amphetamine Confirmation Kit (see Table 2.2). Therefore, it is recommended that an alternate method be used for confirmation (i.e., TLC, GC, GC/MS).

Cross-reactivity is also seen with labetalol, a drug used for the treatment of hypertension (22). Structurally, labetalol is a secondary amine that metabolizes to form a 1-methyl-3-phenyl-propylamino side chain. This side chain reacts with the amphetamine antibody. Benzathine (N,N'-dibenzylethylenediamine) penicillin V, a bacterial antibiotic, has been found to interfere with the EMIT–st Amphetamine Assay (23). One must also consider that additional compounds may readily react with the oxidant (e.g., ketones or glucose in diabetic urine samples), thus leaving the sympathomimetic amine intact to react with the antibody.

2.2. Barbiturates

Barbiturates are a class of drugs capable of producing a state of depression of the central nervous system (CNS) resembling normal sleep, hypnosis, drowsiness, or sedation (18, 19). Barbiturates are used as therapy for many disease states, such as epilepsy, gastrointestinal disorders, hypertension, and asthma, and in preparation for minor medical and dental surgical procedures. Barbiturates in combination with other sedatives markedly increase CNS depression with possible lethal effects. The signs and symptoms of barbiturate abuse are those resulting from depression of the central nervous and cardiovascular systems. A marked tolerance and physical dependence develops to all barbiturates, and withdrawal from their use is one of the most dangerous withdrawal states. Street terms for barbiturates are "goof balls," "yellow jackets," "reds," and "rainbows."

The combination of urea and malonic acid is malonylurea or barbituric acid. The major barbiturate derivatives result by making appropriate substitutions at the CH_2 position of the molecule. Barbiturates are absorbed in the small intestine and metabolized for the most part in the liver. Metabolism includes oxidation, dealkylation, and ring cleavage.

Table 2.2. Cross-reactivity with EMIT Amphetamine Assay

Interference **Eliminated** *by EMIT Confirmation Procedure*

Drugs Containing Ephedrine

Azpan	Marax	Rynatuss
Bronkaid	Marax DF	Tedral
Bronkolixir	Midrane CG	Tedral SA
Bronkotabs	Pazo	T-E-P
Bronkotuss	Primatine P and M	Theozine
Derma Medicone HC	Quadrinal	Vatronol
Efed II	Quelidrine	Wyanoids
Hydroxy Compound	Quibron Plus	

Drugs Containing Pseudoephedrine

Actifed	Dexbron	Respaire SR
Afrinol	Dimacol	Robitussin PE
Ambenyl D	Dimetane Dx	Rondec
Anafed	Disophrol	Ru-Tuss
Atrohist L.A.	Dorcol	Ryna Liquid
Benadryl Decongestant	Drixoral	Ryna C and CX
Beta-Phed	Fedahist	Sine-Aid
Brexin L.A.	Fedrazil	Sinufed
Bromfed	Formula 44D and 44M	Sinutab
Cardec	Guiafed	Sudafed
Chlorafed	Histalet	Teldrin
Chlor-Trimeton	Isoclor	Triafed
Codimal L.A.	Kronofed	Triafed C
Congess Jr. and Sr.	Mediquell	Trinalin
Congestac	Novafed	Tripodrine
Contac	Novahistine Cold and	Tussend
CoTylenol Cold	DMX	Tylenol
Medication	Nucofed	Ursinus
Day Care	Nyquil	Viro-Med
Deconamine	Pediacare 2 and 3	Zephrex
Deconsal	Probahist	Zephrex-L.A.
Detussin		

Drugs Containing Phenylpropanolamine

A.R.M.	Bayer Children's Cold	CCP Tablets
Acutrim	Tabs and Cough	Cheracol Plus
Alka-Seltzer Plus	Syrup	Chexit
Allerest	Borigasic	Children's CoTylenol
Appedrine	Bromphen	Children's Hold

Table 2.2. (*Continued*)

Drugs Containing Phenylpropanolamine (Continued)

Codimal	E.N.T.	Rhinolar
Comtrex	Entex	Robitussin CF
Congespirin	Fiogesic	Ru-Tuss
Contac	4-Way	Sinarest
Control	Head & Chest	Sine-Off
Coricidin	Histalet	Sinubid
Coryban-D	Histamic	Sinulin
Cremacoat 3 and 4	Hycomine	S-T Forte
Decongestant Elixir and	Kronohist	St. Joseph's Cold
Expectorant	Naldecon Dx and Ex	Tablets for Children
Decontabs	Nolamine	Sucrets
Demazin	Ornacol	Super Odrinex
Dexatrim	Ornade	Tavist-D
Dietac	Ornex	Triaminic
Dieutrim	Poly-Histine	Triaminicin
Dimetane-DC	Prolamine	Triaminicol
Dimetapp	Propagest	Trind
Dristan	Pyrroxate	Tuss-Ade
DuraTap PD	Quadrahist	Tussagesic
Dura-Vent	Rhindecon	Tuss-Ornade
Dura-Vent/A		

Drugs Containing Phenylephrine

Albatussin	DuraTap PD	Phenergan VC
Atrohist Sprinkle	Dura-Vent DA	Prefrin Liquifilm
Bromphen	E.N.T.	Protid
Codimal	Entex	Quadrahist
Colrex	Extendryl	Quelidrin
Comhist	4-Way Nasal Spray	Robitussin NR
Congespirin	Histalet	Ru-Tuss
Coricidin	Histaminic	S-T Forte
Coryban-D	Histaspan D and Plus	Sinarest
Dallergy	Histor D	Sinex
Deconsal Sprinkle	Hycomine	Singlet
Decontabs	Korigesic	Spec-T
Dimetane	Naldecon	Tussar DM
Donatussin	Neo-Synephrine	Tussirex
Donatussin DC	NoStril	Tympagesic
Dristan	Novahistine	Rynatan
Dristan AF	P-V-Tussin	Rynatuss
Duo-Medihaler	Pediacof	

Table 2.2. (*Continued*)

---

*Interference **Not** Eliminated by EMIT Confirmation Procedure*

Drugs Containing Amphetamine/d-Amphetamine/Methamphetamine

Obetrol	Dexedrine	Desoxyn

Drugs Containing Isoxsuprine

Vasodilan

Drugs Containing Nylidrin

Arlidin Tablets

Drugs Containing Phenmetrazine

Preludin

Drugs Containing Phentermine

Adipex P	Oby-Trim	Teramine
Fastin	Span R/D	

Source: Ref. (21).

The classification of barbiturates is based on the duration of action: ultra-short-acting, short-acting, intermediate-acting, and long-acting. Long-acting barbiturates are excreted with a large percentage of unchanged drug, whereas short-acting barbiturates are extensively metabolized and excreted with a lower percentage of unchanged drug. In general, the shorter the duration of action, the greater the lipid solubility and percent protein bound. The half-life is variable and not directly related to duration of action; secobarbital may not be detectable after 24 hr, whereas phenobarbital may be detectable for 2–3 weeks after ingestion.

The EMIT Barbiturate Assays (7, 8) screen for the drug class barbiturates. The urine assay low calibrator contains 0.3 mg/L secobarbital and the medium calibrator contains 1.0 mg/L secobarbital. The following drugs at the indicated concentrations cause positive results: phenobarbital (3.0 mg/L), allylbarbital (3.0 mg/L), amobarbital (2.0 mg/L), pentobarbital (1.0 mg/L), butabarbital (1.0 mg/L), and talbutal (2.0 mg/L). The serum low and medium calibrators contain 3.0 mg/L and 6.0 mg/L secobarbital, respectively. (See package insert for levels of serum barbiturates detected.) The sample requirement is serum or plasma using heparin, EDTA, or oxalate as an anticoagulant. This assay is not recommended for whole blood.

Glutethimide in toxic concentrations greater than 25 mg/L has been found to elicit a positive response in the urine barbiturate assay. (See Table 2.3 for over-the-counter and prescription drugs that contain bar-

Table 2.3. Drugs Producing a Positive Response with EMIT Barbiturate Assays

Drugs Containing Amobarbital (Urine and Serum Assays)
Tuinal

Drugs Containing Butabarbital (Urine and Serum)
Butisol
Pyridium Plus
Quibron Plus

Drugs Containing Butalbital (Serum)

A.3.C. Compound	Buff-A Comp	Phrenilin
Amaphen	Esgic	Repan
Anoquan	Fiorinal	Sedapap
Axotal	G-1, 2, and 3	Triad
Bucet	Pacaps	Two-Dyne

Drugs Containing Pentobarbital (Urine and Serum)
Nembutal
Wigraine PB

Drugs Containing Phenobarbital (Urine and Serum)

Antispasmodic Capsules	Chardonna-2	Primatine P Tablets
Antrocol	Levsin	Quadrinal
Acro-Lase Plus	Levsinex	Solfoton
Azpan	Mudrane	T-E-P
Bronkolixir	Mudrane CG	Theofedral
Bronkotabs	Phazyme PB	

Drugs Containing Secobarbital (Urine and Serum)

Seconal	Tuinal

Drugs Containing Talbutal (Urine and Serum)
Lotusate

Source: Ref. (21).

biturates.) Differentiation and confirmation of barbiturates should be performed by an alternate method (i.e., TLC, GC, GC/MS, or HPLC).

2.3. Benzodiazepines

Benzodiazepines are a class of drugs prescribed for use as muscle relaxants, anticonvulsants, hypnotics, and anxiolytics (18, 19). In addition to the drugs available in the United States, Europe markets six other generic benzodiazepines. Chlordiazepoxide, oxazepam, lorazepam, clorazepate, halazepam, alprazolam, and prazepam are anxiolytics. Clorazepam is an anticonvulsant, and flurazepam and temazepam are hypnotics. Diazepam is prescribed for all of the above conditions. Abuse of benzodiazepines is common owing to widespread prescription use.

Benzodiazepines seldom produce either tolerance or physical dependence. However, they are psychologically addictive, thus leading to overeruse. Chronic use may result in some moderate tolerance and dependence. Lethal overdoses rarely occur, because the drugs rarely cause significant circulatory or respiratory depression. The most frequent side effect is drowsiness.

Metabolism occurs in the liver via desmethylation or conjugation. The half-lives of benzodiazepines are variable, with active metabolites resulting in half-lives of up to 100 hr. The drug is rapidly distributed in the body, but excretion is slow. Oxazepam is the final metabolite for many of the parent compounds. Detection can occur several days after large doses of the parent drug. An individual using a drug for years may show detectable benzodiazepine levels in urine for weeks.

The EMIT benzodiazepine assay (9, 10) screens for the drug class, although some benzodiazepines, such as triazolam, may not be detected. Sheep antibody to a diazepam derivative is utilized in the urine assay. Oxazepam is present in the urine assay low and medium calibrators in concentrations of 0.3 mg/L and 1.0 mg/L, respectively. The following drugs may be detected: chlordiazepoxide and lorazepam at levels of 3.0 mg/L and clonazepam, demoxepam, desalkylflurazepam, N-desmethyl-diazepam, diazepam, flunitrazepam, flurazepam, and nitrazepam at levels of 2.0 mg/L. The EMIT serum toxicology low and medium calibrators contain diazepam at levels of 0.3 mg/L and 2.0 mg/L, respectively. Serum or plasma using EDTA, heparin, or oxalate may be used. Whole blood may not be used in the serum toxicology assay. (See the package insert for levels of benzodiazepines detected. See Table 2.4 for over-the-counter and prescription drugs containing benzodiazepines.) The structure of the benzodiazepine appears to be related to its reactivity in EMIT assays (24, 25). No interferences from structurally unrelated compounds are known to exist.

2.4. Cannabinoids

Cannabis sativa is an Indian hemp plant (18, 19) that contains a psychoactive component identified as delta-9-tetrahydrocannabinol (Δ^9-THC). "Marijuana" is the common term for the leaves and flowers. Hashish is the resin extract of the flowering plant. Street terminology includes "weed," "pot," "grass," "dope," "joint," and "roach." It is widely

Table 2.4. Drugs Producing a Positive Response with EMIT Benzodiazepine Assays

Drugs Containing Chlordiazepoxide (Urine and Serum Asays)

Clipoxide	Librium	Menrium
Librax	Limbitrol	Sk-Lygen
Libritabs		

Drugs Containing Clonazepam (Urine and Serum)
Clonopin

Drugs Containing Diazepam (Urine and Serum)

Valium	Valrelease

Drugs Containing Flurazepam (Urine and Serum)
Dalmane

Drugs Containing Lorazepam (Urine and Serum)
Ativan

Drugs Containing Oxazepam (Urine and Serum)
Serax

Drugs Containing Prazepam (Serum)
Centrax

Drugs Containing Temazepam (Serum)
Restoril

Source: Ref. (21).

used as a recreational drug to alter mood and perception. It is commonly abused by smoking and occasionally abused by oral ingestion. Tolerance develops, but no physical dependence is seen. Some adverse reactions include tachycardia, paranoia and, in rare instances, acute psychotic reactions.

Delta-9-THC enters the bloodstream minutes after smoking or 1.5–3 hr after oral ingestion. The drug is rapidly transformed by liver enzymes to several metabolites. The major urinary metabolite of Δ^9-THC is 11-nor-Δ^9-THC-9-carboxylic acid. Owing to marked variation in absorption, distribution, and excretion patterns, it is not possible to correlate psychoactive effects and urine metabolite levels. The drug is highly lipophilic. Screening can only indicate past use; detection can occur for 1–3 days after occasional use or for several weeks after chronic use.

The EMIT cannabinoid assays (6) screen for several cannabinoid metabolites. Assays are available with either a 20-ng or a 100-ng cutoff, with protocol modifications available for a 50-ng cutoff. Sheep antibody to 11-nor-Δ^8-THC-9-carboxylic acid protein conjugate is used. The cannabinoid 100-ng assay calibrators contain 100 and 400 ng/mL of 11-nor-Δ^8-THC-9-carboxylic acid. The 11-nor-Δ^8-THC-9-carboxylic acid employed in the calibrator has been shown to produce results equivalent to that of the metabolite 11-nor-Δ^9-THC-9-carboxylic acid. Other metabolites producing a positive response are 8B, 11-dihydroxy-Δ^9-THC, 8B-hydroxy-Δ^9-THC, 11-hydroxy-Δ^8-THC, and 11-hydroxy-Δ^9-THC at a level of 200 ng/mL and Δ^9-THC at a level of 400 ng/mL.

Cone et al. (26) recently demonstrated that under ordinary environmental conditions passive inhalation does not produce positive results in the 100-ng assay. The rumor has been promulgated that the high concentration of melanin metabolites in the urine of black people may cause positive results with cannabinoid assays. However, a study conducted by El Sohly et al. at the University of Mississippi concluded that this was not the case. Compounds chemically related to the major melanin compounds found in the urine of dark-skinned people showed no cross-reactivity with the assay, and urines from volunteers also gave negative results (27).

2.5. Cocaine and Cocaine Metabolites

Cocaine is an alkaloid of the plant *Erythroxylon coca* (18, 19). Cocaine is a CNS stimulant. Clinically, it is used as a local anesthetic and vasoconstrictor. Recreational abuse has soared in the last decade. It is ingested by snorting, smoking ("free basing"), or intravenous injection. It has strong psychological addiction because of the rapid onset and disappearance of its effects of euphoria and stimulation. High doses may result in paranoid disturbances, delusions of omnipotence, tachycardia, and respiratory depression. Street terms include "coke," "snow," "white lady," "C," "lady," "crack," and "girl."

The chemical structure of cocaine is benzoylmethylecgonine. After oral absorption, cocaine is rapidly inactivated to ecgonine methyl ester by plasma and liver esterases. Owing to this rapid hydrolysis, the half-life of cocaine is 2–4 hr. Cocaine accounts for less than 10% of the drug found in urine. The two main metabolites, benzoylecgonine (35–54%) and ecgonine methyl ester (32–49%), are detected in urine. Benzoylecgonine can be detected within 4 hr of inhalation and for up to 48 hr at levels greater than 1.0 mg/L (11).

The EMIT cocaine metabolite assay (11) contains sheep antibodies specific for benzoylecgonine. The low and medium calibrators contain 0.3 and 3.0 mg/L benzoylecgonine, respectively. Ecgonine and cocaine may also be detected but only at significantly higher levels: 5.0 mg/L and 25 mg/L, respectively.

Because degradation of the drug has been noted if the sample is refrigerated for more than 3 days, it is recommended that samples be stored frozen. Several herbal teas have been found to contain cocaine in quantities sufficient to elicit detectable levels of the cocaine metabolite (27), but the Drug Enforcement Agency has had them withdrawn from the market.

2.6. Methadone

Methadone is a synthetic narcotic analgesic used in the treatment of heroin addiction (18, 19). Commercial methadone is available as Amidone or Dolophine. Structurally different from morphine, methadone has analgesic properties equivalent to morphine. Tolerance and addiction are known to occur. Abuse may result in respiratory depression. Oral dosages are absorbed in the gastrointestinal tract, unlike morphine or heroin. The half-life is 8–18 hr, permitting less frequent administration than with heroin. Methadone may be detected in urine for approximately 3 days after ingestion.

The EMIT methadone assay (12) is designed to detect methadone in urine. The low and medium calibrators contain 0.3 and 1.0 mg/L methadone, respectively. This assay does not detect metabolites of the long-acting form of methadone, 1-α-acetylmethadol (LAAM), in concentrations that would be found in the urine of patients on LAAM therapy.

High concentrations of doxylamine (found in several over-the-counter cold medications) will produce a positive result with the methadone assay (see Table 2.5). Diphenhydramine in overdose situations has also been cited as giving false positive results; however, diphenhydramine sensitivity varies from lot to lot (28–31).

Table 2.5. Drugs Producing a Positive Response with EMIT Methadone Assay

Drugs Containing Doxylamine
Contac Severe Cold Formula
Cremacoat 4
Formula 44
Nyquil Nighttime Colds Medicine
Unisom Nighttime Sleep Aid

Source: Ref. (21).

2.7. Methaqualone

Methaqualone has sedative, hypnotic, anticonvulsant, antitussive, and weak antihistaminic effects (18, 19). It is similar in action to barbiturates. No longer marketed as Quaalude in the United States, it is still available on foreign markets as Sopor, Parest, and Quaalude. Nicknames include quaas, ludes, sopes, or sopors. Both tolerance and physical dependence may develop. Adverse effects may include dry mouth, tongue discoloration, anorexia, and peripheral neuropathy. Overdose may result in restlessness, hypertonia, and seizures, and death may occur from respiratory arrest or pulmonary edema.

The absorption of methaqualone [2-methyl-3-*o*-tolyl-4(3H)-quinazolinone] is rapid, with peak plasma concentrations being reached in 1.5–2 hr. It is extensively metabolized in the liver; 12 hydroxylated metabolites and an *N*-oxide metabolite have been identified. Less than 1% is excreted in the urine as the unchanged drug. Chronic therapy has been found to induce hepatic microsomal enzymes. The drug may be detected for 2 weeks after use and has a half-life of 6–18 hr.

The EMIT assay utilizes sheep antibody to methaqualone. The methaqualone assay (13) is specific for the parent compound and three major urinary metabolites. The low and medium calibrators contain 0.3 mg/L and 1.5 mg/L methaqualone, respectively. The assay also detects mecloqualone (a Schedule I drug marketed in Europe and South Africa), 3'-hydroxy-methaqualone, and 4'-hydroxy-methaqualone at a level of 1.0 mg/L and 2'-hydroxymethyl-methaqualone at a level of 5.0 mg/L.

2.8. Opiates

Opiates are a class of narcotic drugs manifesting sedative, mood-altering, and analgesic properties (18, 19). They include the naturally occurring alkaloids from opium, morphine, and codeine; semisynthetic opiates, such as heroin (diacetylmorphine), oxycodone, and hydromorphone; and synthetic opiates, such as meperidine. The drugs may be taken by snorting, subcutaneous or intravenous injection, or smoking. Opiate abuse results in marked tolerance and physical and psychological dependence. Opiate overdose exhibits the triad of stupor or coma, respiratory depression, and pinpoint pupils. Death from overdose may be a result of respiratory depression or acute pulmonary edema.

Morphine is rapidly absorbed, with peak levels from an oral dose occurring after 15–60 min and from an injection occurring after 15 min. It is metabolized extensively in the liver. Only 2–12% is excreted as unchanged drug, and 60–80% of the conjugated metabolites is excreted in the urine. Quantitatively, the most important metabolite is morphine-3-glucuronide. The half-life of morphine is 1.7–4.5 hr. Heroin is metabolized to monoacetylmorphine and further to morphine. Codeine is metabolized to codeine-6-glucuronide, with 10–15% being metabolized to morphine and norcodeine by demethylation. In general, opiates are metabolized by hydrolysis, oxidation, N-dealkylation, and hepatic conjugation with glucuronic acid. Opiates may be detected in urine up to 2 days after use (32).

The EMIT Opiate Assay (14) utilizes a sheep antibody to a morphine derivative. The low and medium calibrators contain 0.3 mg/L and 1.0 mg/L morphine, respectively. Codeine and hydrocodone are detected at a level of 1.0 mg/L; hydromorphone, levorphanol, and morphine-3-glucuronide at a level of 3.0 mg/L; and oxycodone at a level of 50 mg/L. (See Table 2.6 for over-the-counter and prescription drugs that contain opiates.)

Meperidine in high concentrations may elicit a positive response. Ingestion of poppy seeds may also produce a positive response, if the seeds contain opiates left from an incomplete washing process (27, 33–36).

2.9. Phencyclidine

Phencyclidine, 1-(1-phenylcyclohexyl)piperidine (PCP), was first used as an experimental general anesthetic under the trade name Sernyl (18, 19). Advantages appeared to be its nonnarcotic, nonbarbituric, and nonrespiratory depressive properties. However, PCP was found to have psychotic side effects profound enough to warrant removal from the market for human use. During the late 1960s, it was available as a veterinary anesthetic, Sernylan, and during this time, drug abuse began to appear. Combativeness, nystagmus, catatonia, convulsions, coma, hallucinations, schizophrenia, depression, flashbacks, and social withdrawal are some of the symptoms observed in PCP abuse. PCP may be inhaled, injected, or taken orally. PCP is synthesized with relative ease. Several analogs that have similar hallucinogenic effects have been synthesized as well. It is unknown whether physical dependence results from the use of PCP.

PCP is well absorbed following all routes of administration. The maximum concentration is reached 5–15 min after smoking or 2 hr after oral ingestion. The drug undergoes oxidation and conjugation in the body with only 10% of the dose being excreted in its unchanged form. The half-life ranges from 8–55 hr, and the drug is more readily excreted in acidic urine. There is a high degree of tissue distribution, accounting for the drug being excreted for days or weeks after ingestion.

The EMIT assay (15) contains a sheep antibody to a phencyclidine

Table 2.6. Drugs Producing a Positive Response with EMIT Opiate Assays

Drugs Containing Codeine

Acetaco	G-Z and G-3 Capsulen	Ryna C and CX
Ambenyl Cough Syrup	Guiatuss A-C Syrup	Sedapap #3
Bromanyl Expectorant	Iophen-C Liquid	Stopayne Capsules and
Bromphen DC	Naldecon CX	Syrup
Expectorant	Novahistine DH	Triafed-C Expectorant
Codalan	Novahistine	Tussar SF
Codimal PH	Expectorant	Tussar 2
Decongestant	Nucofed	Tussi-Organidin
Expectorant	Pediacof Cough Syrup	Tussirex
Decongestant AT	Poly-Histine-CS	Compounds with
Dimetane DC	Robitussin AC	codeine in trade name
Dolprn #3	Robitussin DAC	

Drugs Containing Hydrocodone
Tussionex

Drugs Containing Hydromorphone
Dilaudid

Drugs Containing Levorphanol
Levo-Dromoran

Drugs Containing Morphine

Duramorph	MSIR Tablets	Roxanol
MaContin	RMS Suppositories	Roxanol SR

Drugs Containing Oxycodone

Percocet	Percodan	Tylox Capsules

Source: Ref. (21).

derivative. The low and medium calibrators contain 75 and 400 ng/mL phencyclidine, respectively. Several of the PCP metabolites and analogs elicit a positive result (see Table 2.7). Ketamine, PCA, and PCC do not cross-react. High concentrations of thoridiazine, dextromethorphan, and chlorpromazine may produce positive results (37).

2.10. Propoxyphene

Propoxyphene is a synthetic narcotic analgesic (18, 19) structurally similar to methadone and is marketed as Darvon compound, Darvocet, and Darvon-N (propoxyphene capsylate). Drug abuse usually results from an excessive use of prescribed drug. The ingestion of large doses results in

Table 2.7. Compounds Producing a Positive Response with EMIT PCP Assays

Compound	Level (μg/mL)
1-(1-Phenylcyclohexyl)morpholine (PCM)	1.0
1-(1-Phenylcyclohexyl)pyrrolidine (PCPy)	1.0
1-[1-(2-Thienyl)-cyclohexyl]piperidine (TCP)	1.0
1-[1-(2-Thienyl)-cyclohexyl]pyrrolidine (TCPy)	1.0
4-Phenyl-4-piperidinocyclohexanol	2.0
N,N-Diethyl-1-phenylcyclohexylamine (PCDE)	3.0
1-(4-Hydroxypiperidino)phenylcyclohexane	3.0
1-[1-(2-Thienyl)-cyclohexyl]morpholine (TCM)	5.0

Source: Ref. (15).

tolerance and physical dependence. Toxicity and eventual death result from pulmonary edema and respiratory depression.

Propoxyphene[(S) - α - [2 - (dimethylamino) - 1 - methylethyl] - α - phenylbenzene ethanol propanoate] is pharmacologically related to the opiates, and the half-life ranges from 2–6 hr. Propoxyphene may be detected in urine for 2 days after ingestion.

The EMIT assay (16) utilizes a sheep antibody to propoxyphene. The low and medium calibrators contain 0.3 and 1.0 mg/L propoxyphene, respectively. The assay is designed to detect propoxyphene, propoxyphene salts, and norpropoxyphene (the major urinary metabolite) with equal sensitivity. (See Table 2.8 for over-the-counter drugs producing positive responses.)

Methadone may interfere if present in toxic levels greater than 100 mg/L. Urinary levels in methadone maintenance patients are less than 50 mg/L. Dextromethorphan at levels greater than 100 mg/L may also give a

Table 2.8. Drugs Producing a Positive Response with
EMIT Propoxyphene Assay

Drugs Containing Propoxyphene
Darvon
Dolene AP-65
Dolene Capsules
Lorcet
SK-66
Wygesic

Source: Ref. (21)

positive response. However, these concentrations are not found in urine following therapeutic doses.

2.11. Tricyclic Antidepressants

The two major actions of tricyclic antidepressants (TCAs) are antidepressant and anticholinergic (18, 19). In toxic overdoses, CNS effects include hallucinations and confusion prior to coma. Anticholinergic effects include dry mouth and dilated pupils. Cardiotoxic effects, such as a prolonged QRS interval and cardiac arrythmias, hypotension, and pulmonary edema, lead to increased mortality. Sudden death has been reported up to 6 days after overdose.

The structure of TCAs consists of a central ring bounded by two benzene rings. The tricyclics differ by radicals attached to the central ring. The parent compound is a tertiary amine, whereas its major metabolite is a secondary amine. TCAs are rapidly absorbed from the gastrointestinal tract. Their metabolism in the liver consists of demethylation, hydroxylation, and conjugation. The half-life is 24–76 hr but may be longer in overdose.

The EMIT TCA assay for serum is designed to detect toxic levels of TCAs (17). Amitriptyline (Elavil), nortriptyline (Aventyl), imipramine (Tofranil), desipramine (Pertofrane), and related metabolites are the primary tricyclics detected by this serum assay. A sheep antibody to desipramine is used. The serum calibrator and positive control contain 300 and 1000 ng/mL nortriptyline, respectively. (Refer to Table 2.9 for the sensitivities of TCAs and their metabolites and to Table 2.10 for drugs containing TCAs.)

High therapeutic doses of chlorpromazine (200–300 ng/mL) (17) as well as other phenothiazines (38) may give a positive response. Diphenhydramine at levels greater than 120 ng/mL may also elicit a positive response (39).

3. CONCLUDING REMARKS

Each immunoassay must be characterized in terms of sensitivity and cross-reactivity. Many studies are conducted to verify that substances other than the drug or drug class do not cause a positive result. Any substance thought to interfere is studied extensively, and if interference is found, this is noted in the package insert. The quality assurance of an assay includes testing of drug-free samples as well as samples containing

Table 2.9. Drugs Producing a Positive Response with EMIT Serum Toxicology TCA Assay

Compound	Level (ng/mL)
Amitriptyline	240–360
Desipramine	200–300
Imipramine	200–300
10-Hydroxyamitriptyline	<1000
10-Hydroxynortriptyline	<1250
2-Hydroxydesipramine	<1250
2-Hydroxyimipramine	<750
Amoxapine	<500
Clomipramine	<500
Doxepin	<500
Protriptyline	<500
Trimipramine	<600

Source: Ref. (17).

Table 2.10. Drugs Containing TCAs

Drugs Containing Amitriptyline

Domical	Etrafon	Saroten
Elavil	Lentizol	Triavil
Endep	Limbitrol	Tryptizol

Drugs Containing Nortriptyline

Allegron	Aventyl	Pamelor

Drugs Containing Imipramine

Berkomine	Janimine	SK-Pramine
Imavate	Presamine	Tofranil

Drugs Containing Desipramine

Norpramin	Pertofrane

Drugs Containing Doxepin

Adapin	Curatin	Sinequan

Drugs Containing Protriptyline

Concordin	Vivactil

Source: Ref. (21).

varying drug and/or metabolite concentrations, performing assays on a variety of instruments, and testing at independent laboratories (40).

Enzyme immunoassays should be employed as initial drug testing methodologies. Positive results should be confirmed by a second methodology, such as thin-layer chromatography, high performance liquid chromatography, gas chromatography, or gas chromatography/mass spectrometry (GC/MS).

In drug testing, a false positive occurs when a drug is reported to be present but is not actually present. This may be due to operator error, or it may be due to the presence of a substance mistakenly identified as the drug being analyzed. Such a chemical false positive may be due to cross-reactivity. The antibody used in the assay may bind to a compound that has a similar chemical structure to the drug being tested (1). Quality assurance will help remove operator error; confirmatory testing will uncover chemical false positives.

In some cases, the alternative test used to confirm a positive initial test will not confirm the presence of the drug owing to differences in methodology. The confirmation of some drugs requires hydrolysis before confirmation by a chromatographic method. There also may be a difference in the sensitivity of the assays or in the assay's ability to detect metabolites of the drug. In these situations, an EMIT positive may be labeled as a false positive when the term "unconfirmed positive" may be more correct.

The potential interference of nonsteroidal anti-inflammatory drugs with some EMIT drug abuse assays was recognized in early 1986 (41). This potential interference was suggested by the inability to confirm some EMIT positive results by GC/MS. The interference was subsequently found to affect only a few assays, and this interference was eliminated by reformulating the assays affected (27). Recently, Podkowik et al. (42) reported that metabolites of nonsteroidal anti-inflammatory drugs may cause low recovery of substances on GC/MS by reacting with the derivatizing reagent. This raises questions as to the degree of positive interference from nonsteroidal anti-inflammatories seen with EMIT assays.

It is important to keep up to date with product information with regard to sensitivities and cross-reactivities. As shown in the sections on individual assays, there are many prescription and over-the-counter medications that may give positive results. Because of the legal implications of drug screening results, it is important to be aware of all information on cross-reactivity, other interferences, and medications and compounds that contain the drug being tested. In all cases, positive results by EMIT assays should be confirmed by a second methodology.

REFERENCES

1. *Syva Monitor,* 4(2) 1986, Syva Co., Palo Alto, CA.

2. Emit-d.a.u. Drug Abuse Urine Assays package insert, Syva Co., Palo Alto, CA, 1981.

3. J. Turner and J. Chang, *AACC, tdm-t,* 5(6) 1983.

4. *Syva Monitor,* 5(2) 1987, Syva Co., Palo Alto, CA.

5. Emit-d.a.u. Amphetamine Assay package insert, Syva Co., Palo Alto, CA, September 1984.

6. Emit-d.a.u. Cannabinoid 100 ng Assay package insert, Syva Co., Palo Alto, CA, April 1986.

7. Emit-d.a.u. Barbiturate Assay package insert, Syva Co., Palo Alto, CA, August 1984.

8. Emit-tox Serum Barbiturate Assay package insert, Syva Co., Palo Alto, CA, May 1986.

9. Emit-d.a.u. Benzodiazepine Assay package insert, Syva Co., Palo Alto, CA, August 1985.

10. Emit-tox Serum Benzodiazepine Assay package insert, Syva Co., Palo Alto, CA, 1984.

11. Emit-d.a.u. Cocaine Metabolite Assay package insert, Syva Co., Palo Alto, CA, September 1984.

12. Emit-d.a.u. Methadone Assay package insert, Syva Co., Palo Alto, CA, August 1984.

13. Emit-d.a.u. Methaqualone Assay package insert, Syva Co., Palo Alto, CA, June 1985.

14. Emit-d.a.u. Opiate Assay package insert, Syva Co., Palo Alto, CA, August 1984.

15. Emit-d.a.u. Phencyclidine Assay package insert, Syva Co., Palo Alto, CA, August 1984.

16. Emit-d.a.u. Propoxyphene Assay package insert, Syva Co., Palo Alto, CA, March 1985.

17. Emit-tox Serum Tricyclic Antidepressants Assay package insert, Syva Co., Palo Alto, CA, 1984.

18. A. Goth, *Medical Pharmacology,* 10th ed., Mosby, St. Louis, MO, 1981.

19. L. M. Haddad and J. F. Winchester, *Clinical Management of Poisoning and Drug Overdose,* Saunders, Philadelphia, 1983.

20. Amphetamine Confirmation Kit package insert, Syva Co., Palo Alto, CA, 1984.

21. Syva Co. communication.

22. F. S. Apple, M. K. Googins, S. Kastner, K. Nevala, S. Edmondson, and J. Kloss, *Clin. Chem., 31*(7), 1250–1251 (1985).

23. N. R. Badcock and G. D. Zoanetti, *Clin. Chem.*, *33*(6), 1080 (1987).

24. R. D. Budd, *Clin. Toxicol.*, *18*, 643–655 (1981).

25. M. Manchon, M. F. Verdier, P. Pallud, A. Vialle, F. Beseme, and J. Bienvenu, *J. Anal. Toxicol.*, *9*, 209–212 (1985).

26. E. J. Cone, R. E. Johnson, W. D. Darwin, and D. Yousefnejad et al., *J. Anal. Toxicol.*, *11*, 89–96 (1987).

27. *Syva News Report*, Summer 1986, Syva Co., Palo Alto, CA.

28. C. E. Pippenger and R. S. Galen, *Diagn. Med.*, Sept./Oct. 1983, p. 84.

29. M. J. Kelner, *Clin. Chem.*, *30*(8), 1430 (1984).

30. L. A. Ajel, *Clin. Chem.*, *31*(2), 340–341 (1985).

31. C. E. Pippenger, K. Joyce, G. Erenberg, and F. Van Lente, *Clin. Chem.*, *30*(6), 1031 (1984).

32. H. M. Vandenberghe, S. J. Soldin, and S. M. MacLeod, *AACC-TDM*, Nov. 1982.

33. R. E. Struempler, *J. Anal. Toxicol.*, *11*, 97–99 (1987).

34. A. M. Zebelman, B. L. Troyer, G. L. Randall, and J. D. Batjer, *J. Anal. Toxicol.*, *11*, 131–132 (1987).

35. L. W. Hayes, W. G. Krasselt, and P. A. Mueggler, *Clin. Chem.*, *33*(6), 806–808 (1987).

36. B. C. Pettitt, Jr., S. M. Dyszel, and L. V. S. Hood, *Clin. Chem.*, *33*(7), 1251–1252 (1987).

37. R. L. Hawks and C. N. Chiang, *Urine Testing for Drugs of Abuse*, NIDA Research Monograph No. 73 DHHS Publ. No. (ADM) 87-1481, Washington, DC, 1986.

38. T. J. Schroeder, J. J. Tasset, E. J. Otten, and J. R. Hedges, *J. Anal. Toxicol.*, *10*, 221–224 (1986).

39. A. Sorisky and D. C. Watson, *Clin. Chem.*, *32*(4), 715 (1986).

40. EMIT Drug Abuse Assays: How Accurate Are They?, Syva Co., Palo Alto, CA (1986).

41. *Syva News Report*, Spring 1986, Syva Co., Palo Alto, CA.

42. B. Podkowik, M. L. Smith, and R. O. Pick, *J. Anal. Toxicol*, *11*, 215–218 (1987).

CHAPTER

3

BONDED PHASE EXTRACTION IN ANALYTICAL TOXICOLOGY

MARTHA R. HARKEY

University of California, Davis
Davis, California

An extraction procedure is usually necessary in analytical toxicology to provide a relatively clean sample for analysis (thereby improving selec-

59

tivity and precision) and to concentrate the sample (improving the sensitivity of analysis). However, the initial sample extraction steps are often the most tedious, time-consuming, and variable aspect of analysis. Advances in instrumentation have led to more rapid and selective chromatographic separations, automated sample injection, and computer-assisted data analysis, all serving to improve the speed and precision of analyses; by comparison, there have been relatively few improvements in sample extraction techniques. Thus sample preparation remains the limiting factor for a totally automated analysis and the source of greatest variability in results.

Solid phase extraction materials are becoming more popular in analytical toxicology because they provide a more rapid and selective extraction, require much smaller volumes of organic solvents, and provide a means to automate sample extraction.

1. LIQUID–LIQUID VERSUS SOLID PHASE EXTRACTION TECHNIQUES

Liquid–liquid extraction remains the most commonly used sample preparation technique in analytical toxicology. Drugs are extracted from biological samples by partitioning between two immiscible solvents, often adjusting pH to control ionization and back-extracting at a different pH to improve selectivity and recovery. Differential pH extraction is relatively nonspecific but fairly reproducible and requires no special supplies or instrumentation. Its disadvantages are the time required for multi-step extractions, the expense and disposal considerations for large volumes of organic solvents, and the limited precision resulting from nonselective extraction of other compounds that may interfere with analysis and/or cause variable recovery of the analyte(s). In addition, emulsions are often formed that may require additional steps to produce an adequate separation of phases.

The principle of solid phase extraction is similar to that for liquid–liquid extraction, but involves a partitioning of compounds between a solid and liquid phase rather than between two immiscible solvents. The extracted compound should have a stronger affinity for the solid phase than the sample matrix, yet be easily removed with a small volume of solvent. This two-step extraction principle is summarized in Fig. 3.1 which shows retention when a compound has a higher affinity for the solid phase (S) than the sample matrix (L) and elution when its solubility is greater in the elution solvent (L).

Liquid   Solid

$$X \rightleftharpoons X$$

$$K = \frac{[X]_S}{[X]_L}$$

$K >>> 1 \longrightarrow$ **Retention**

$K <<< 1 \longrightarrow$ **Elution**

Figure 3.1. Distribution of compound (X) between liquid (L) and solid (S) phase. Compound is retained on solid phase when association constant (K) is much greater than 1, and eluted when K is much less than 1.

1.1.   Solid Phase Adsorbents

A number of solid phase adsorbents have been used to extract drugs from biological samples. Materials such as charcoal, alumina, silica, agarose, and hydrophobic resins are relatively nonspecific in extracting drugs from biological matrices, but can be useful in developing extraction procedures for comprehensive toxicological screening for a wide range of drugs. The adsorbents most often used in analytical toxicology are a cross-linked divinylbenzene–polystyrene resin, XAD-2® (1, 2) or diatomaceous earth, Clin-Elut® (3).

1.2.   Bonded Silica Adsorbents

The use of chemically bonded silicas for sample preparation followed their use in high performance liquid chromatography (HPLC) to provide a selective separation of compounds. These materials usually provide a better recovery of drugs from biological matrices and may be more convenient than other procedures used in analytical toxicology (4, 5). Stewart et al. (4) compared the recovery of five chemical classes of drugs (acidic, basic, amphoteric, hydrophobic, and hydrophilic) from water and plasma using six commercially available solid phase adsorbents (XAD-2, Clin-Elut, silica, cyano-bonded silica, and octadecyl-bonded silica from two manufacturers). Extractions with silica were the least reproducible, whereas octadecyl-bonded silica generally produced the best precision and highest recoveries. Interestingly, significant differences in percent recovery were also found between octadecyl-bonded silicas prepared by two different manufacturers. A limitation of this study, however, is that all materials were used according to manufacturers' directions and were not optimized for extraction of particular drugs (4).

In another study comparing extractions of four drugs from urine using liquid–liquid extractions with dichoromethane, XAD-2, Celite® (diatomaceous earth), and octadecyl-bonded silica, no significant differences were found in total recoveries. However, extraction with octadecyl-bonded silica was the preferred method because much less organic solvent was used and the procedure was faster and easier to perform (5).

The disadvantages of using bonded silicas for sample preparation have been the irreproducibility of extraction recoveries for some analytes and the cost of the columns (typically $1.00–2.00 each). Variable recoveries may result from improper conditioning of extraction columns, failure to maintain an optimal pH for extraction, or incomplete extractions due to interactions between the drug and sample matrix. Most of these problems can be eliminated with proper use of these materials, as described in manufacturers' handbooks (6–8).

The cost of extraction columns has not deterred their use significantly, for it is largely offset by the reduced sample preparation time and limited use of organic solvents. These columns can be used more than once; however, recovery from recycled columns varies with the drug extracted and the column used. Before reusing columns, recovery must be validated with different lots of bonded phase to ensure reproducibility and columns should be used only for the predetermined number of samples (typically 3–10).

2. BONDED SILICA EXTRACTION PRINCIPLES

2.1. Chemistry of Bonded Silicas

Many methods are used in the preparation of bonded silicas; however, all rely on the covalent attachment of the bonded phase through the reaction between a siloxane modifier and surface silanols on silica. As shown in Fig. 3.2, the stoichiometry of the reaction is largely determined by the functionality of the modifier. The simplest reaction occurs with monofunctional siloxane derivatives that produce a monolayer of bonded phase. The stoichiometry of these reactions is one to one (modifier to silanol); however, derivatization is never complete. For example, a monomeric modifier with minimal steric hindrance, such as trimethylchlorosilane, leaves approximately 50% unreacted surface silanols (9). Polyfunctional reactions, which are commonly used in the preparation of bonded phases for sample extraction, produce a more complex surface. Because these modifiers tend to polymerize in solution, the surface is covered with a dense coating of bonded phase. The stoichiometry of these

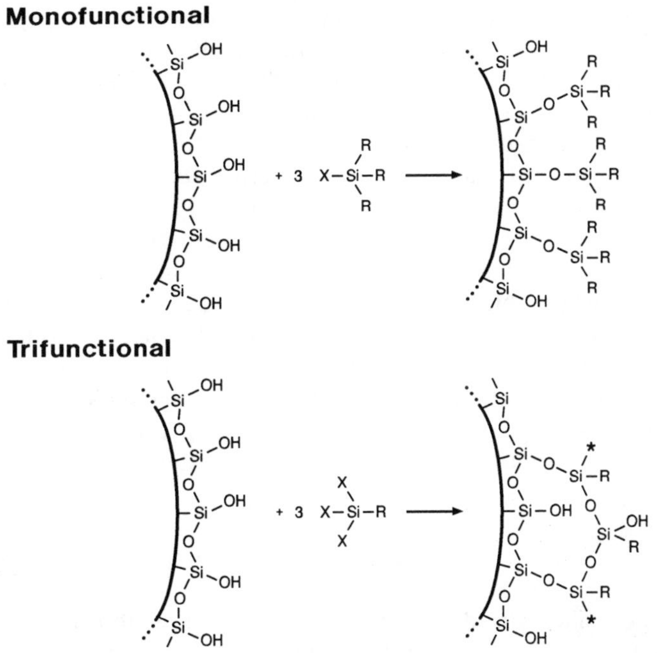

Monofunctional

Trifunctional

Figure 3.2. Preparation of chemically bonded silica by monofunctional and trifunctional derivatization. Hydrolysis of unreacted Si–X groups on trifunctional modifiers results in formation of residual silanols or further polymerization of bonded phase (as noted by ''*'').

reactions may be one modifier for one to two silanols, depending on the degree of polymerization and the accessibility of surface silanols. As shown in Fig. 3.2, polyfunctional reactions may also result in residual silanols on the modifier when unreacted Si–X groups are hydrolyzed (9). Although the stoichiometry suggests that a more complete reaction would occur with monomeric derivatives, all modifications of silica leave a significant number of reactive surface silanols. Because residual silanols alter the retention characteristics of the bonded phase, these are often inactivated by ''end-capping'' with trimethylchlorosilane or similar reagents.

Both monofunctional and polyfunctional alkylsiloxanes have been used in the preparation of bonded silicas, each with advantages or disadvantages depending on their use (10). Because trifunctional modifiers tend to polymerize in solution, both the surface area and the volume of pores are reduced significantly (9). Access to residual surface silanols is also limited, so polymeric bonded silicas are more nonpolar than monomeric derivatives. In general, the chemistry of monomeric bonded phases tends to be more reproducible, whereas polymeric phases are more stable and have a higher percentage of carbon loading. New phases that combine the advantages of both monofunctional and trifunctional derivatization are being synthesized (11). As new reagents for the preparation of bonded silicas are evaluated, more selective materials for both chromatography and sample preparation should be available.

Variable recovery of some analytes may be avoided if the chemistry of bonded phase preparation is known. For example, hydrochloric acid is produced when chlorosiloxane derivatives are used as modifiers. Depending on the manufacturing process and quality control, residual acidity of the phase may vary considerably from lot to lot, which can lead to irreproducible extractions of weakly acidic compounds. This problem may be resolved by conditioning the bonded phase with an acidic buffer prior to adding sample, thus ensuring a consistent pH for each lot of materials.

2.2. Physicochemical Properties of Bonded Silicas

The bonded silicas used in sample preparation are 40-μm irregular particles with 60-Å pores. These larger particles (as compared to the 3–10 μm materials used in analytical columns) require much less pressure to move solvents through the packing material, so extractions can be performed on a small manifold with laboratory vacuum or pressure. In some cases, a small bench-top centrifuge has been used by placing the extraction columns in a centrifuge tube.

Although bonded silicas are not stable above pH 7–7.5 (9), extremes in pH (up to 13) have been used in specific applications for sample preparation. The bonded phase withstands higher pH in sample extraction because the phase is in contact with solutions for a very short period of time, unlike HPLC, where the material is in continuous contact with the mobile phase. In addition, the phase rapidly equilibrates with solvents, so the pH is easily restored.

The capacity of bonded silicas is proportional to the surface area and concentration of bonded phase, as well as the length of the bonded alkyl group (9). Capacity becomes important in sample preparation to determine the amount of material that can be effectively retained during extraction. Most nonpolar phases have a capacity of 1–5% of the mass of the bonded phase; thus 1–5 mg of analyte can be retained on 100 mg of material (6). Although a capacity of milligrams provides an excess for extracting the usual nanogram to microgram quantities of drugs in analytical toxicology, capacity can be reduced by the nonselective extraction of sample matrix components. Also, because the analyte is in equilibrium between the bonded phase and solvent (see Fig. 3.1), capacity is decreased by passing large volumes of solution through the extraction column (12).

2.3. Phases Available for Sample Extraction

A variety of bonded phases have been developed for use in sample preparation and are classified according to functional group as nonpolar, polar, and ion exchange (Fig. 3.3). Nonpolar phases are frequently used in analytical toxicology because most drugs are easily extracted from aqueous solutions by hydrophobic phases. In addition, experience with reversed phase liquid chromatography is directly transferable to sample prepara-

NON-POLAR

C18	Octadecyl	$-Si-C_{18}H_{37}$
C8	Octyl	$-Si-C_8H_{17}$
C2	Ethyl	$-Si-C_2H_5$
CH	Cyclohexyl	$-Si- \hexagon$
PH	Phenyl	$-Si- \bigcirc$

POLAR

CN	Cyanopropyl	$-Si-CH_2CH_2CH_2CN$
2OH	Diol	$-Si-CH_2CH_2CH_2OCH_2CH-CH_2$ $\qquad\qquad\qquad\qquad OH \quad OH$
SI	Silica	$-Si-OH$
NH₂	Aminopropyl	$-Si-CH_2CH_2CH_2NH_2$
PSA	N-propylethylenediamine	$-Si-CH_2CH_2CH_2NCH_2CH_2NH_2$ $\qquad\qquad\qquad\qquad H$

ION EXCHANGE

SCX	Benzenesulfonylpropyl	$-Si-CH_2CH_2CH_2-\bigcirc-SO_3^{\ominus}$
PRS	Sulfonylpropyl	$-Si-CH_2CH_2CH_2-SO_3^{\ominus}$
CBA	Carboxymethyl	$-Si-CH_2COO^{\ominus}$
DEA	Diethylaminopropyl	$-Si-CH_2CH_2CH_2\overset{\oplus}{N}(CH_2CH_3)_2$ $\qquad\qquad\qquad\qquad H$
SAX	Trimethylaminopropyl	$-Si-CH_2CH_2CH_2\overset{\oplus}{N}-(CH_3)_3$

Figure 3.3. Nonpolar, polar, and ion exchange phases available for sample preparation. (Figure provided by Analytichem International, a subsidiary of Varian Instruments.)

tion with nonpolar phases. Shorter-chain nonpolar phases can be very selective in extraction; however, the selectivity of extraction decreases as the length of the carbon chain increases. Even though octadecyl (C_{18}) is the least selective nonpolar phase, it has been the most often used in analytical toxicology because it extracts virtually any drug, toxicologists are familiar with its characteristics through experience with HPLC, and it is available from all manufacturers of bonded silica.

Polar phases such as cyanopropyl, diol, aminopropyl, or N-propylethylenediamine have been used for a limited number of applications in toxicology. These phases are more selective for extracting compounds with hydroxyl or amine groups, because the primary attraction between analyte and bonded phase is through hydrogen bonding rather than the weaker van der Waals forces associated with nonpolar extractions. Although polar phases are potentially more selective for polar drugs, nonpolar phases are most often used in analytical toxicology because nonpolar extractions are less subject to variability in extraction due to changes in pH or ionic strength. Of the polar phases, cyano has been used most frequently for sample extraction as well as in liquid chromatography.

Strong ion exchange phases are rarely used in sample preparation because the strong attraction between analyte and bonded phase can result in variable recovery. Interestingly, most of the applications have been in extractions of drugs from horse urine. For example, ion exchange columns have been used as an initial clean-up column attached to a nonpolar column. In this dual column extraction, salts are retained by the ion exchange column and analytes are extracted by the second column. The extraction of a broad spectrum of drugs has recently been reported with a strong cation exchange phase, phenylsulfonic acid. In this procedure, 280 drugs are isolated on the bonded phase, then selectively eluted by varying the pH of eluting solutions (13). A strong anion exchange resin has been used for the selective extraction of diethylstilbestrol from liver extracts prior to analysis by gas chromatography/mass spectrometry (GC/MS) (14).

Mixed-function bonded phases have been used to improve selectivity by providing interactions with both a nonpolar and polar phase. An example is the Clean-Screen℗ extraction column, which contains both a hydrophobic and an ionic moiety. This material may be useful in general toxicology screening because drugs with various functional groups can be extracted simultaneously, then selectively eluted with different eluting solvents (15).

2.4. Configuration of Extraction Systems

Bonded silicas are packed either in individual cartridges that can be attached to a syringe or in syringe barrels, as shown in Fig. 3.4. The barrel

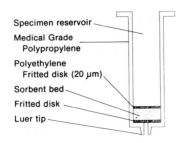

Specimen reservoir

Medical Grade
  Polypropylene

Polyethylene
  Fritted disk (20 μm)

Sorbent bed

Fritted disk

Luer tip

Figure 3.4. Extraction column used for sample preparation (Figure provided by Analytichem International, a subsidiary of Varian Instruments.)

is made from medical grade polypropylene with the bonded material held in place between two polyethylene fritted disks. The most common problem associated with these columns has been solubilization of plasticizers in the frits when strong eluting solvents are used. This usually occurs when acetone is used as the eluting solvent and may be avoided by using a column with stainless steel frits.

Extraction time can be reduced by processing a number of samples simultaneously by inserting the columns into a pressure or vacuum manifold. A typical extraction sequence requires 5 min or less per batch of 10–30 samples (depending on the size of the manifold).

Other configurations have been developed specifically for use with automated extraction devices. These generally consist of cassettes which have 5–10 individual cartridges, each packed with 50–500 mg of bonded silica (e.g., AASP® and Auto Spe-ed™).

3. DEVELOPMENT OF EXTRACTION PROTOCOL

The first step in developing an extraction protocol is choosing the bonded phase. This is largely empirical, but may be based on chromatographic data for the analyte or previous experience with extracting similar compounds. The more tedious steps are determining the wash and elution solutions to optimize selectivity, reproducibility, and sensitivity of extraction.

3.1. Choice of Bonded Phase

Sample extraction with bonded silicas is based on the same principles as liquid chromatography, so the selection of a phase can be determined by chromatographic characteristics of the drug whenever possible. If the drug chromatographs on a reversed phase analytical column, the strength of the mobile phase will give a good approximation of the retention on an extraction column. For example, if the mobile phase is relatively weak in organic composition (e.g., less than 15% organic solvent), a more hydrophobic nonpolar phase will be necessary to retain the drug during sample cleanup. Conversely, if a stronger mobile phase is used (e.g., greater than 50% organic solvent), a relatively weak nonpolar phase can be used for greater selectivity in extraction.

Another consideration is chromatographic peak symmetry. Although "tailing" of chromatographic peaks can be caused by a number of variables, one common cause is the interaction between polar substituents on the drug molecule and surface silanols. This may be eliminated by the addition of triethylamine or nonylamine to the mobile phase. The need for an amine modifier in the mobile phase to improve chromatography suggests that a more polar phase may be used for sample extraction. For example, the cyano phase has both nonpolar and polar characteristics and can be used to extract drugs that have polar functional groups, such as antidepressants (16).

The choice of bonded phase also depends on the sample matrix. For aqueous samples such as serum or urine, nonpolar extraction is often used because nonpolar drugs are easily extracted from an aqueous sample onto a nonpolar phase and recovery depends less on pH and salt concentrations. To improve the selectivity of extraction, polar or weak ion exchange phases may be used; however, these phases often require sample pretreatment (e.g., sample dilution or pH adjustment) to efficiently extract a drug from the sample matrix.

Although most drugs are extracted with the nonpolar C_{18} phase, greater selectivity and sensitivity can be achieved by using a less hydrophobic nonpolar phase. For example, the drug amiodarone is very lipophilic and is well retained on all nonpolar phases. By using the less retentive C_2 phase, fewer serum components or other drugs are extracted, potential interferences are more easily removed with wash solutions, and smaller volumes of eluting solvent are needed to elute amiodarone and its metabolite from the bonded phase (17).

To reduce method development time, a number of phases may be evaluated simultaneously; however, the lipid solubility and polar characteristics of the drug should first be considered. If the drug is very hydrophobic, the initial trial should include C_2 and C_8 phases, as well as C_{18} for comparison. Alternatively, if the drug has polar substituents, a cyano and/or diol phase may be more selective than a C_8 or C_{18}. The C_2 phase has limited utility in analytical toxicology because many drugs are poorly retained and may be eluted during subsequent wash steps.

Multiple extraction columns have been used to improve the selectivity of extraction. Usually the compound of interest is retained on the first column, with most of the extraneous compounds passing through the column. A second extraction column is then attached directly beneath the first column, so that the analyte can be eluted onto the second column for concentration or further purification. Alternatively, sample contaminants may be retained on the first column while the analyte is extracted by the second column. Using multiple extraction columns increases the expense of sample preparation, but can be used for samples that contain multiple interfering compounds or when the analysis requires a very selective extraction.

A final consideration is tailoring the sample extraction to the method of analysis. To decrease total sample preparation time, the eluent from the bonded phase can be injected directly onto a gas or liquid chromatograph. Thus the choice of bonded phase may be critical in providing an eluent that is both compatible with the analytical instrument and contains the drug in the suitable concentration. For example, some detectors used in liquid chromatography are sensitive to refractive index changes and produce spikes or negative deflections with samples containing 100% organic solvent. This variability in detector response can be avoided by using a weaker hydrophobic phase and eluting the drug with a solution more similar to the mobile phase. Also, when sensitivity is a concern, the drug must be eluted in as small a volume as possible. By choosing a weaker hydrophobic phase, smaller volumes of elution solvent can be used, often eliminating evaporation and reconstitution steps.

3.2. Developing the Extraction Sequence

A typical extraction sequence is shown in Fig. 3.5. The bonded silica, or sorbent, is first activated with an organic solvent, such as methanol or acetonitrile, that conditions the bonded phase by wetting surface silanols and bonded functional groups. For 100-mg extraction columns, approximately 1 mL solvent is used and may be added simply by filling the syringe barrel with solvent from a squeeze bottle. This is often referred to as one "column volume." For nonpolar extractions, the conditioning solvent must be hydrophobic in order to interact with the nonpolar functional groups on the bonded silica; however, it must also be miscible with the sample matrix because residual solvent molecules are always retained in the bonded phase. The next step is to remove the organic solvent and condition the phase for sample extraction by adding approximately 1 mL water or buffer. This step prepares the phase for the sample, so a solution similar to the sample matrix is used (e.g., for nonaqueous extractions, an organic solvent is used). After each conditioning step, the vacuum or pressure is released to prevent drying because excessive drying alters the

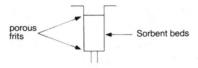

1. Activation of Sorbent
2. Removal of Activation Solvent
3. Application of Sample
4. Removal of Interferences (I)
5. Elution of Concentrated, Purified Analytes (A)

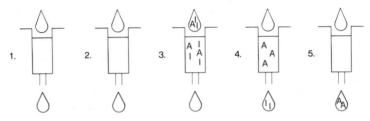

Figure 3.5. A typical extraction sequence for sample preparation. (Figure provided by Analytichem International, a subsidiary of Varian Instruments.)

retention characteristics of the bonded silica and may result in variable recoveries.

After conditioning the bonded phase, the sample is added (usually 0.1–1 mL). Serum or plasma samples are usually added directly, but urine may be diluted with water or buffer to improve extraction from very concentrated samples. When extracting whole blood, the hemolysed sample is filtered or centrifuged to remove cellular debris that can clog the frits in the extraction column or physically interfere with extraction. As the sample passes through the extraction column, components will either bind to the bonded phase or pass through the column unretained, depending on the affinity of the compound for the bonded phase as compared to its solubility in the sample matrix. Ideally, the analyte is selectively extracted by the bonded phase whereas other sample components pass through the column unretained. In practice, however, this is rarely the case; therefore this step is usually followed with a wash step(s).

After the sample is extracted potential interferences may be removed by washing the column with solvents of varying strengths. For most nonpolar phases, water can be used to remove many of the polar constituents of serum or urine without eluting drugs. More nonpolar contaminants may be removed by adding relatively weak solutions of methanol or acetonitrile in water or buffer. The strength of wash solutions may be estimated from the chromatographic conditions (wash solutions have a much lower organic composition than the chromatographic mobile phase). To ensure that no "breakthrough" or loss of analyte occurs during wash steps, the eluent should be saved for analysis.

Finally, the analyte is eluted from the extraction column. The elution solvent for nonpolar phases is usually methanol, acetonitrile, or a mixture of organic solvent with water or buffer. Depending on the concentration of the analyte and the method of analysis, this solution may be analyzed directly without evaporation and reconstitution. To improve sensitivity the analyte should be eluted in as small a volume as possible, which typically concentrates the samples five- to tenfold. However, the volume of eluting solvent must be sufficient to pass through all pores and interstitial spaces in the bonded phase. This volume is governed by the particle size and average pore size of the bonded silica. For example, at least 100 μL is required for 100 mg of the bonded silica used in sample preparation (40μm particle with 60-Å pores). For most applications, two to five times this volume may be needed for complete recovery.

A general protocol to develop a method for drug extraction with a nonpolar bonded phase is summarized as follows:

1. Condition the column with 1 mL (or one "column volume") methanol or acetonitrile, followed by 1 mL water or buffer. A buffer may be used in extracting weak acids or bases to ensure that the drug is nonionized.

2. Add solution of drug standard (in the same concentration range as is expected in biological samples) to the column. Collect eluate.

3. Wash column with one or more column volumes wash solution(s). To determine appropriate wash solutions, start with one column volume water or buffer. If the drug is retained on the column, progressively stronger solutions (e.g., 5%, 10%, 15% methanol or acetonitrile) may be used. Determine the strongest solution that will remove sample contaminants without eluting the drug, then choose a slightly weaker solution as wash solvent to ensure reproducibility of extraction with lot to lot variability in bonded material. In each trial, the eluate is retained to check for "breakthrough" (partial elution of drug with wash solutions).

4. Identify appropriate elution solvent for compatability with analytical method. (If necessary, the drug may be eluted with one solvent, then evaporated and reconstituted in another solvent for analysis; however, this adds time and additional expense to the analysis.)

5. Elute the drug with elution solvent such as methanol, acetonitrile,

or a mixture of organic solvent with water or buffer. The strength of elution solvent can be estimated from the strength of wash solutions.

6. Repeat extraction sequence using blank sample and samples spiked with the expected concentrations of the drug(s). Observe efficiency of wash solutions in removing contaminants in blank sample, and the reproducibility and linearity of recovery of drug from spiked solutions. At this point, modifications of wash and/or elution solutions may be necessary to provide a cleaner sample for analysis or to improve recovery.

3.3. Potential Problems

Variable recovery of drugs can result from lot to lot differences in the bonded phase. Even though manufacturers maintain quality control for these materials, their procedures may not detect minor variations in bonded phases that can alter recoveries for a particular drug. Some phases are now available that are quality controlled for particular drugs and marketed under that drug name (e.g., THC phase). Recovery problems can usually be overcome by developing an extraction protocol that is not sensitive to minor changes in the bonded phases (e.g., using weaker wash solutions to minimize breakthrough and either increasing the volume or using stronger elution solutions to improve recovery). In addition, at least three different lots of materials should be evaluated during method development and only one supplier of bonded phase should be used after the procedure is finalized. Because bonded silicas supplied by different manufacturers may have different retention characteristics, changing manufacturers requires validation of the new material before it can be routinely used.

Reproducibility may also be affected by the flow rate through the column (12). When pressure or vacuum manifolds are used for sample preparation, higher flow rates through the column do not usually alter recovery; however, for less strongly retained compounds slower flow rates may be needed. If "breakthrough" is a problem, flow rates can be decreased by attaching a two-way valve between the column and the manifold to adjust the flow.

Secondary interactions between drugs and bonded silica can result in variable recoveries owing to incomplete elution of analyte. With nonpolar extractions, secondary interactions are most often due to interactions between polar functions on the drug and residual silanols. In this case, amine modifiers such as triethylamine or nonylamine can be added to the eluting solution to improve recovery. When a polar phase is used in sample preparation, nonpolar interactions between the drug and alkyl spacers on the bonded phase can improve extraction by enhancing retention. However, the eluting solution must contain organic solvent such as methanol or acetonitrile to disrupt nonpolar secondary interactions.

Matrix effects, which result from the interactions between the sample matrix and the bonded phase or between the analyte and sample components, can also alter recovery. Matrix-bonded phase interactions are rarely a problem in analytical toxicology owing to the excess capacity of most bonded phases for extracted drugs; however, these interactions result in the coextraction of nonspecific compounds that may interfere with or prolong the analysis. By using a more selective bonded phase or modifying wash and/or elution solutions, extraction of matrix components can be reduced.

A more common matrix problem is the interaction between a drug and serum components. Matrix problems should be suspected whenever the recovery of a drug from a spiked sample is less than the recovery from a pure solution. These matrix effects are most often due to drug–protein binding and may be overcome by precipitating proteins with acetonitrile, trichloroacetic acid, or other reagents and diluting prior to extraction, by using protein denaturing reagents, by changing the pH, or by increasing salt concentration. For example, recovery from a serum sample spiked with amiodarone was much less than that observed with standard solutions. This drug has a high affinity for serum proteins and is greater than 95% protein bound. In this case, recovery from serum samples was restored by pretreating serum samples with 2 M sodium acetate prior to extraction (17).

Other matrix effects, caused by ionic strength, pH, and presence of particulate matter in the sample, are easily resolved. Variable recoveries due to ionic strength can be eliminated by diluting the sample. Effects of pH on recovery are easily overcome by buffering the sample and bonded phase or adjusting the pH of the sample with acid or base. Cellular debris can be removed from whole blood with an inert filter attached directly to the extraction column or by centrifuging the sample prior to extraction.

4. NOVEL APPLICATIONS AND NEW DEVELOPMENTS

There have been a number of new applications and developments for the use of bonded silicas in sample preparation. These include combining the use of bonded silicas for sample extraction with other techniques such as immunoassay to improve selectivity of analysis and the production of special phases for more selective extractions. The most significant advancement for routine toxicology applications, however, has been the introduction of semiautomated or fully automated systems for sample preparation that can be interfaced directly to a liquid or gas chromatograph.

4.1. Novel Applications

Bonded silica extraction has recently been coupled with immunoassay to improve the specificity of analysis for tricyclic antidepressants (18, 19). In this procedure, the problem with cross-reactivity of drug metabolites was resolved by using a bonded phase that had much stronger affinity for the parent drug than for hydroxylated metabolites. After the bonded phase was conditioned, the sample was added to the extraction column, metabolites were eluted from the column with a wash solution, and then the parent drug was eluted for analysis by immunoassay. Extraction of the sample prior to analysis by immunoassay eliminated the interference from the hydroxylated metabolites, thereby providing a more accurate analysis for the parent drugs.

Another interesting application is the use of off-line bonded silica extraction to facilitate LC/MS, LC/NMR, and LC/FTIR (20). The use of reversed phase liquid chromatography provides an ideal separation method for many compounds; however, subsequent analysis by mass spectrometry (MS), nuclear magnetic resonance (NMR), and Fourier transform infrared spectroscopy (FTIR) is limited by the water–organic mixtures used in chromatography. To overcome this problem, compound(s) are extracted by a solid phase extraction column as they elute from the liquid chromatograph, then eluted from the extraction columns with a solvent appropriate for subsequent analysis by MS, NMR, or FTIR.

4.2. Special Phasess

The synthesis of special phases for highly selective extraction represents an important development in the use of these materials for sample preparation. For example, derivatives of boronic acid have been covalently attached to either agarose (21) or silica (22), providing a very selective extraction mechanism for *cis*-diol-containing compounds, such as catecholamines. The principle for this selective extraction is as follows: the phase is ionized with alkaline solution prior to extraction, *cis*-diol-containing compounds are selectively extracted through covalent interaction with the ionized phase, and then the compounds are eluted with an acidic solution that suppresses ionization of the phase, thereby disrupting the covalent attachment of analyte. Although similar phases have not yet been developed for the analysis of specific drugs, this approach offers potential for very selective extractions.

4.3. Automated Extraction Systems

The solid phase extraction systems described previously have reduced sample preparation time considerably; however, handling large numbers of samples can still be tedious owing to the size limitations of extraction systems. A number of approaches have been used to automate sample extraction with bonded silicas partially or fully. The first procedure is the use of column switching techniques for sample preparation prior to analysis by HPLC. Serum samples are injected directly onto a precolumn that is plumbed through a switching valve to the analytical column. Early eluting polar components are eluted to waste with a weak mobile phase solution while the analyte is retained on the precolumn (these systems require either two separate isocratic pumps or a gradient system). The valve is then switched so that the mobile phase flows through the precolumn, eluting the analyte onto the analytical column. Depending on the system, the valve may remain in this position for a short period to allow the analytes to be eluted onto the analytical column, then switched back to the original position, trapping late eluting peaks (which can be eluted to waste) on the precolumn. These systems have been successfully used in therapeutic drug monitoring to provide quantitation for a variety of drugs with no sample preparation (23).

A similar system has recently been described for automated sample preparation by dialysis and trace enrichment (24). The extraction system consists of autosampler/injector, two dilutors, dialyzer, trace enrichment cartridge (15 mg C_{18} bonded phase), injection valve, and liquid chromatograph. The sample is mixed with internal standard and injected into the donor channel of the dialyzer by the autosampler/injector, then held static. The dialysate is pumped through the recipient channel of the dialyzer and through the cartridge by the dilutor. The injection valve is then switched and the analytes are eluted from the cartridge onto the liquid chromatograph with the mobile phase. The dialyzer is purged and the bonded phase cartridge reequilibrated. The rate-limiting step for this system is the chromatographic time, because each sample is chromatographed immediately after extraction.

Another approach has been to automate sample preparation by using robots and solid phase extraction columns (25). The robot performs all the necessary steps from initial sample handling, conditioning the extraction column, adding the sample, washing the column, and then finally positioning the extraction column over the autosampler and eluting the sample directly into the autosampler vial for analysis.

A similar approach has been developed for use with the MilliLab™ Workstation, which consists of a transport system and fluidics module (26). The transport system contains sample test tubes, individual sample filters, Sep-Pak® cartridges, and a probe that handles samples, dispenses liquids, and performs the filtration and extraction steps. Totally automated extraction and analysis by HPLC is possible either with an adapter for filling WISP® vials or a valve-loop injector that injects the eluate from the Sep-Pak directly onto a liquid chromatograph.

A semiautomated sample extraction/injection system (AASP) has been developed for off-line sample extraction and on-line injection onto a chromatograph (27). The AASP cassettes, each containing 10 individual cartridges packed with 50-mg bonded silica, are conditioned and the sample added manually at a specially designed sample preparation station. The cassettes are then placed in the AASP for automatic elution and injection of the samples onto a liquid chromatograph or similar instrument. With the use of a 10-port injection valve, solutions can be pumped through the cassette to remove sample contaminants with a purge solution, then the valve is switched and the mobile phase flows through the cartridge eluting the sample onto the analytical column (as with column switching). The valve is held in this position for a specified period of time, then switched back to the load position, trapping late eluting peaks on the AASP cartridge. To automate the manual steps of sample preparation a robot can be used to prepare the cassettes before loading onto the AASP (28). The system may also be fully automated with the use of a small computer-controller, appropriate switching valves, and an autosampler to inject the sample onto the AASP cassettes (29).

A similar instrument (Auto Spe-ed) has been recently introduced that provides fully automated sample preparation (30). Samples are added to test tubes that are then placed in a tray and loaded into the Auto Spe-ed. Cassettes (each containing five cartridges with 500-mg bonded phase), solvent reservoirs, disposable pipette tips, and collection tubes are all contained within the instrument. The extraction proceeds according to preprogrammed instructions: the cassettes are conditioned, samples are added, and cassettes are washed with appropriate solutions, dried under nitrogen, and then eluted into collection tubes or interfaced directly to a chromatograph. Analyses for benzoylecgonine and delta-9-carboxytetrahydrocannabinol in urine have recently been reported with this system (31).

5. CONCLUSIONS

The design and manufacture of instruments to automate sample extraction and the development of new and/or more reproducible materials for extraction offer significant improvements in sample preparation for analytical toxicology. The potential for automated extraction systems is evident by the number of instruments outlined in Section 4.3, many of which can be interfaced to an analytical instrument to automate the analysis fully. A number of new materials are also being introduced for sample preparation, including new or mixed phases bonded to either silica or polystyrene-based materials (32). Finally, as the differences in the chemical and structural properties of bonded silicas are better understood (33), more reproducible materials should become available for routine use in sample preparation.

ACKNOWLEDGMENTS

The author would like to thank Dr. Michael Burke of the Department of Chemistry, University of Arizona and Dr. Terry Sheehan of Varian Instruments for discussions on the chemistry of bonded silicas; Dr. Ron Majors of EM Science for information on new procedures for automating sample preparation; and Drs. Tom Grove and Vince Perna for permission to reproduce procedures developed in the Research Department of SmithKline Bio-Science Laboratories.

APPENDIX: EXTRACTION OF DRUGS WITH BONDED PHASES

1. Tocainide and Mexiletine from Serum or Plasma (Ref. 34)

Sample Preparation (C_8 Bond-Elut®, 100 mg)

1. Condition extraction column by inserting into vacuum manifold and washing with 1 mL 50:50 (acetonitrile:methanol), followed by 2 mL distilled water.
2. Add 0.05 mL internal standard solution (levobunolol, 30 mg/L), followed by 1 mL sample, standard, or control.

3. Wash extraction column twice with 1 mL distilled water, followed by 1 mL 10% acetonitrile in distilled water.

4. Place collection tubes in manifold and elute with 1.5 mL 20:20:60 (acetonitrile: methanol: 100 mM phosphate buffer, pH 4.9).

5. Inject 6 μL onto liquid chromatograph.

Chromatographic Conditions

Analytical Column: Perkin-Elmer C_8, (3 μm, 3 cm cartridge)

Mobile Phase: 15:15:70 (acetonitrile: methanol: phosphate buffer, 100 mM, pH 4.9)

Flow Rate: 0.5 mL/min

Detector: UV at 209 nm

2. Amiodarone and N-Desethylamiodarone from Serum or Plasma (Ref. 17)

Sample Preparation (C_2 Bond-Elut, 100 mg)

1. Pretreat the samples prior to extraction by pipetting 0.5 mL of sample, standard, or control into labeled test tubes. Add 0.3 mL 2 M sodium acetate and 0.01 mL internal standard (L8040,* 200μg/mL).

2. Condition extraction column by inserting into vacuum manifold and washing with 1 mL methanol, followed by 1 mL distilled water.

3. Add pretreated sample to column and apply vacuum.

4. Wash with 1 mL distilled water, followed by 1 mL 50% methanol in water, then 1 mL 50% acetonitrile in water.

5. Place collection tubes in manifold and elute with 1 mL 60:40 (acetonitrile:methanol with 2 mM nonylamine).

6. Inject 6 μL onto liquid chromatograph.

Chromatographic Conditions

Analytical Column: Perkin-Elmer C_8 (3 μm, 3 cm cartridge)

Mobile Phase: 600 mL acetonitrile, 290 mL methanol, 130 mL distilled water, 0.36 mL nonylamine

Flow Rate: 0.5 mL/min

Detector: UV at 254 nm

*Obtained from Sanofi Center for Research, Paris, France.

3. Diazepam and Its Metabolites from Serum, Plasma, or Whole Blood (Ref. 35)

Sample Preparation (C_{18} Bond-Elut, 100 mg)

1. Condition extraction column by inserting into vacuum manifold and washing with one column volume methanol, followed by one column volume distilled water.
2. Prepare column for extraction by adding 0.1 mL 0.1 M sodium borate buffer (pH 9.5) and 0.05 mL heparin (1000 units/mL).
3. Add 0.05 or 0.1 mL sample, standard, or control, followed by 0.01 mL internal standard (methylnitrazepam, 10 mg/L).
4. Wash twice with one column volume water, followed by 0.05 mL methanol.
5. Place collection tubes in manifold and add 0.02 mL methanol. Equilibrate for 2 min before eluting. Repeat with 0.1 mL methanol, then leave vacuum on for 30 s to elute sample completely.
6. Dry the eluate under nitrogen, reconstitute in 0.025 mL mobile phase, and inject 0.01 mL onto liquid chromatograph.

Chromatographic Conditions

Analytical Column: Technicon Fast-LC C-8 (5 μm, 15 cm)
Mobile Phase: 53:46:1 (methanol: 2 mM monobasic potassium phosphate: acetonitrile)
Flow Rate: 1.3 mL/min
Column Temperature: 35°C
Detector: UV at 240 nm

4. Cyclosporine from Whole Blood (Ref. 36)

Sample Preparation (C_{18} Bond-Elut, 100 mg)

1. Pretreat the samples prior to extraction by pipetting 2 mL internal standard (cyclosporine D, 0.2 mg/L in 30:70, acetonitrile: distilled water) into labeled test tubes. Add 1 mL sample, standard or control to each tube. Vortex and centrifuge 5 min at 500 × g.
2. Insert extraction column in vacuum manifold, attach a 5-mL reservoir to extraction column, and condition with 2 mL ethanol followed by 2 mL distilled water.
3. Add hemolyzed sample to reservoir and apply vacuum.
4. Wash with 4 mL wash solution (50% acetonitrile in distilled water).

5. Place collection tubes in manifold and elute with 0.3 mL ethanol.
6. Add 0.2 mL distilled water and 0.5 mL hexane to each tube.
7. Vortex, then centrifuge 1 min at $500 \times g$.
8. Aspirate off hexane layer, pour lower ethanol layer into vials. Inject 0.1 mL onto liquid chromatograph.

Chromatographic Conditions

Analytical Column: Perkin Elmer C_{18} (3 μm, 3 cm cartridge) with Rheodyne 2-μm filter

Mobile Phase: 63:37 (acetonitrile:distilled water)

Flow Rate: 1 mL/min

Temperature: 70°C

Detector: UV at 210 nm

5. Cyclosporine Automated Extraction from Whole Blood (Ref. 37)

Sample Preparation (C_8 AASP cartridge)

1. Pretreat the samples prior to extraction by pipetting 1.5 mL internal standard (cyclosporine D, 0.2 mg/L in 50:50, acetonitrile:5% $ZnSO_4$ in distilled water) into labeled test tubes. Add 0.5 mL sample, standard, or control to each tube. Vortex and centrifuge 2 min at 500 × g.
2. Insert cartridge into pressure manifold and condition cartridge with 1 mL acetonitrile, followed by 0.5 mL distilled water.
3. Add whole blood supernatant.
4. Wash with 1 mL wash solvent (2:3, acetonitrile:distilled water), then insert cartridge into AASP.
5. Purge column with 40 strokes purge solution (7:2:11, acetonitrile, methanol, distilled water) preinjection and 20 strokes postinjection.
6. Set valve reset at 2.5 min, run time at 14 min, and cycle time at 14.5 min.

Chromatographic Conditions

Analytical Column: Varian C-8 (5 μm, 15 cm) with Brownlee C-8, 3 cm guard column

Mobile Phase: 53:20:27 (acetonitrile:methanol:distilled water)

Flow Rate: 1.5 mL/min

Temperature: 70°C

Detector: UV at 210 nm

6. Acidic and Neutral Drugs from Whole Blood (Ref. 38)

Sample Preparation (C$_{18}$ Bond-Elut 100 mg)

1. Pretreat samples prior to extraction by pipetting 1 mL sample, standard, or control into 15-mL centrifuge tubes. Add 10 mL 0.1 M KH$_2$PO$_4$ (pH 3.5) and 1 mL internal standard (metharbital, 20 mg/L or 5-methylphenyl-5-phenylhydantoin, 5 mg/L). Mix thoroughly and centrifuge at low speed for 2 min.

2. Insert extraction column into vacuum manifold, attach reservoir to extraction column, and condition with two column volumes of methanol, followed by two column volumes distilled water.

3. Add diluted blood to reservoir and apply vacuum.

4. Wash reservoir twice with distilled water, then remove from extraction column.

5. Wash extraction column twice with distilled water, then leave vacuum on for 5–10 min to dry. Turn vacuum off.

6. Place collection tubes in manifold, add 0.1 mL acetonitrile, let stand 1 min, then elute with vacuum. Repeat with 0.1 mL acetonitrile, then leave vacuum on to evaporate to dryness (approximately 5 min). Evaporation is not necessary if analyzing by HPLC.

7. Reconstitute with 0.05 mL acetonitrile.

8. Inject 10 μL onto HPLC or 2 μL onto GC.

Note: This procedure was developed for whole blood, but may also be used for urine, serum, or plasma.

Chromatographic Conditions

HPLC

Analytical Column: Beckman C$_{18}$ (5 μm, 25 cm)

Mobile Phase: 15:23:62 (methanol: acetonitrile: 0.1 M KH$_2$PO$_4$, pH 3.5)

Flow Rate: 2 mL/min

Temperature: 50°C

Detector: UV at 195 nm

GC/NPD

Analytical Column: DB-5, 30 m (0.25 μm ID, 0.25 μm film thickness)

Injector Temperature: 250°C

Detector Temperature: 300°C

Temperature Program: 110–to 250°C at 12°/min, final hold time 8 min

Gas Flow: 2 mL/min H_2, 175 mL/min air, 30 mL/minN_2 (make-up)

Detector: Varian 6000 TSD (range 10^{-12})

7. Delta-9-carboxytetrahydrocannabinol from Urine (Ref. 39)

Sample Preparation (Narc℠-1, Carboxy Methyl Ester Column)

1. Hydrolyze sample by pipetting 3 mL urine into 12-mL screw-top tube. Add 3 mL HPLC grade water and 0.3 mL 10 N KOH. Vortex briefly, then heat for 15 min in a 60°C water bath. Cool to room temperature, then adjust pH to between 4 and 6 with approximately 0.4 mL glacial acetic acid.

2. Insert extraction column into vacuum manifold and condition column with one column volume methanol, followed by one column volume 0.05 M phosphoric acid (pH 2.5).

3. With vacuum off, add 2 mL 0.05 M phosphoric acid. Attach a 15-mL reservoir to the column, then add the hydrolyzed urine. Apply vacuum (at 5–8 in. Hg) until the column is dry. Remove reservoir.

4. Wash extraction column slowly with 2 mL 40:60 (acetonitrile: 0.1 N HCl). Dry column under vacuum for 1 min. Turn vacuum off and add 0.5 mL hexane. Apply vacuum to aspirate hexane, then dry column under vacuum (25 in. Hg) for 5 min.

5. Place collection tubes in manifold and elute under low vacuum (5 in. Hg) with 1 mL 50:50 (ethyl acetate:hexane).

6. To derivatize prior to GC/MS, collect eluate in 2 mL reaction vials containing 0.02 mL dimethylformamide. Evaporate under nitrogen, then add 0.04 mL bis(trimethylsilyl)trifluorcacetamide with 1% trimethylchlorosilane. Heat capped vials in dry bath at 70°C for 50 min. Cool and inject onto GC/MS.

Chromatographic Conditions

Analytical Column: HP-1, 12 m (0.2 mm ID, 0.33 μm film thickness)

Injector Temperature: 275°C

Oven Temperature: 270°C (isothermal)

Capillary Direct Interface Temperature: 275°C

Gas Flow: 0.8 mL/min He

Detector: 5790B Mass Selective Detector

MS Parameters: Dwell time 100 msec. SIM of m/z 371, 473, 488 for carboxy THC. SIM of m/z 491 for deuterated carboxy THC

8. Acidic, Basic, and Neutral Drugs from Urine (Ref. 15)

Sample Preparation (Clean-Screen)

1. Pretreat sample prior to extraction by pipetting 5 mL sample, standard, or control into test tube. Add 2 mL 0.1 M phosphate buffer (pH 6.0) and adjust to pH 5.5–6.5 with 0.1 N NaOH or 1 N acetic acid.

2. Insert extraction column into vacuum manifold and condition column with 1 mL methanol followed by 1 mL 0.1 M phosphate buffer (pH 6.0).

3. Attach an 8-mL fritted reservoir to the top of extraction column and add urine sample. Elute slowly with vacuum (3–5 in. Hg).

4. Wash with 1 mL 0.1 M phosphate buffer followed by 0.5 mL 1 N acetic acid.

5. Dry column for 5 min with vacuum, then wash with 1 mL hexane.

6. Place collection tubes in manifold. Elute acidic and neutral drugs with four aliquots of 1 mL methylene chloride.

7. Evaporate eluate under nitrogen at 30–40°C, then reconstitute in 0.1 mL ethyl acetate and inject 1–2 μL onto GC.

8. To elute basic drugs, wash columns with 1 mL methanol, then place collection tubes in manifold. Elute with 2 mL basic methanol (2% NH_4OH in methanol).

9. Add 3 mL distilled water and 0.2–0.3 mL chloroform to eluate. Vortex 15 seconds and inject 1–2 μL chloroform layer onto GC.

REFERENCES

1. J. M. Fujimoto and R. I. H. Wang, *Toxicol. Appl. Pharmacol.*, *16*, 186 (1970).

2. M. L. Bastos, D. Jukofsky, E. Saffer, M. Chedekel, and S. J. Mulé, *J. Chromatogr.*, *71*, 549 (1972).

3. B. Newton and R. F. Foery, *J. Anal. Toxicol.*, *8*, 129 (1984).

4. J. T. Stewart, T. S. Reeves, and I. L. Honigberg, *Anal. Lett.*, *17*, 1811 (1984).

5. P. M. Hyde, *J. Anal. Toxicol.*, *9*, 269 (1985).

6. K. C. Van Horne, Ed., *Sorbent Extraction Technology*, Analytichem International, Inc., Harbor City, CA, 1985.

7. *Baker-10 SPE Applications Guide*, Vol. 1, J. T. Baker Chemical Company, Phillipsburg, NJ, 1982.

8. *Baker-10 SPE Applications Guide*, Vol. 2, J. T. Baker Chemical Company, Philipsburg, NJ, 1984.

9. K. K. Unger, N. Becker, and P. Roumeliotis, *J. Chromatogr., 125*, 115 (1976).

10. S. A. Wise and W. E. May, *Anal. Chem., 55*, 1479 (1983).

11. R. D. Golding, A. J. Barry, and M. F. Burke, *J. Chromatogr., 384*, 105 (1987).

12. B. A. Bidlingmeyer, *LC, Liq. Chromatogr. HPLC Mag., 2*, 578 (1984).

13. S. Barker, T. McDonald, T. Aranas, and C. Short, Proceedings of the Seventh International Conference of Racing Analysts and Veterinarians, Louisville, KY, April 10 (1988).

14. T. Covey, G. Maylin, and J. Henion, *Biomed. Mass Spectrom., 12*, 274 (1985).

15. Clean-Screen℗ Applications Note, Worldwide Monitoring Corporation, Morrisville, PA.

16. G. L. Lensmeyer and M. A. Evenson, *Clin. Chem., 30*, 1774 (1984).

17. C. Cano, M. Harkey, and T. Grove, unpublished procedure from SmithKline Bio-Science Laboratories.

18. S. Pankey, C. Collins, and A. Jaklitsch, *Clin. Chem., 31*, 909 (1985).

19. S. Pankey, A. Izutsu, C. Collins, A. Jaklitsch, M. Pirio, C. Lin, and M. Hu, *Clin. Chem., 31*, 909 (1985).

20. C. D. Wilcox and R. M. Phelan, *J. Chromatogr. Sci., 24*, 130 (1986).

21. S. Higa, T. Suzuki, A. Hayashi, I. Tsuge, and Y. Yamamura, *Anal. Biochem., 77*, 18 (1977).

22. A. H. B. Wu and T. G. Gornet, *Clin Chem., 31*, 298 (1985).

23. M. Wrenger, M. Oellerich, E. Raude, and M. Riedmann, *Fresenius Z. Anal. Chem. 324*, 350 (1986); Application Notes, Hewlett-Packard Company, Analytical Group, Palo Alto, CA.

24. D. C. Turnell and J. D. H. Cooper, *J.Chromatogr., 395*, 613 (1987).

25. G. Shoenhard, R. Schmidt, L. Kosobud, and K. Smykowski, "Robotic Assays for Drugs in Animal and Human Plasma," in G. L. Hawk and J. R. Strimaitis, Eds., *Advances in Laboratory Automation Robotics 1984*. Zymark Corporation, Inc., Hopkinton, MA, 1984, p. 61.

26. G. J. Fallick, *J. Anal. Purification*, March, p. 24 (1987); *Applications Notes*, Millipore Corporation, Waters Chromatography Division, Milford, MA.

27. J. C. Pearce, J. A. Jelly, K. A. Fernandes, W. J. Leavens, and R. D. McDowall, *J. Chromatogr., 353*, 371 (1986).

28. J. C. Pearce, M. P. Allan, and R. D. McDowall, "Robotics in Drug Analysis," in E. Reid, B. Scales, and I. D. Wilson, Eds., *Methodological Surveys in Biochemistry and Analysis, Vol 16, Bioactive Analytes, Including CNS Drugs, Peptides, and Enantiomers*, Plenum Press, New York, 1986, p. 293.

29. K. C. Van Horne, presented at the Pittsburgh Conference and Exposition, Atlantic City, NJ, March, 1984.

30. R. J. Badesso and M. J. Fetner, presented at the Pittsburgh Conference and Exposition, Atlantic City, NJ, March, 1987.

31. D. E. Nadig and R. J. Badesso, presented at the Pittsburgh Conference and Exposition, New Orleans, LA, February, 1988.

32. Applications Note, Interaction Chemicals, Inc., Mountain View, CA.

33. J. Köhler, D. B. Chase, R. D. Farlee, A. J. Vega and J. J. Kirkland, *J. Chromatogr., 352,* 275 (1986).

34. M. Tsunokawa, M. Harkey, and T. Grove, unpublished procedure from SmithKline Bio-Science Laboratories.

35. S. N. Rao, A. K. Dhar, H. Kutt, and M. Okamoto, *J. Chromatogr., 231,* 341 (1982).

36. P. M. Kabra and J. H. Wall, *J. Liq. Chromatogr., 10,* 477 (1987).

37. P. M. Kabra, J. H. Wall, and P. Dimson, *Clin. Chem., 33,* 2272 (1987).

38. W. J. Phillips, modified from presentation to California Association of Toxicologists, February, 1984.

39. R. E. Kiser, H. E. Ramsden, J. M. Patterson, and D. Law, *Clin. Chem., 32,* 1115 (1986).

CHAPTER

4

GAS CHROMATOGRAPHY/MASS SPECTROMETRY USING TABLETOP INSTRUMENTS

DALE G. DEUTSCH

State University of New York at Stony Brook
Stony Brook, New York

1. HISTORICAL PERSPECTIVE

The applicability of gas chromatography–mass spectrometry (GC/MS) for use in the clinical toxicology laboratory was demonstrated in the early 1970s. One of the earliest clinical applications was by Althaus et al. (1) who, in 1970, identified propoxyphene (Darvon) and metabolites as the agent responsible for coma in a patient admitted to a local hospital. This group from the Massachusetts Institute of Technology assembled a computer-searchable library containing 304 mass spectra of drugs, metabolites, normal constituents, and common sample contaminants. Using a Perkin-Elmer 990 gas chromatograph coupled to an Hitachi RMU 6L mass spectrometer interfaced to an IBM 1800 data acquisition and control system, they summarized their data, in 1974, from more than 600 patient samples (2).

Law et al. (3) in 1971 at the National Institutes of Health developed a library containing 58 commonly used drugs and a semiautomated search routine to help the analyst. This group employed a LKB-9000 GC/MS coupled to a DEC PDP-81 computer.

Finkle et al. (4) in 1972 assembled tables of electron impact (EI) mass spectra for 133 drugs of abuse using a Finnigan Model 3000D GC/MS with a Hewlett-Packard 2000C Computer. The data were presented in both an alphabetical index and base peak index, which allowed manual as well as computer search library identification. In a subsequent report the library was enlarged to include EI and chemical ionization (CI) mass spectra for 450 drugs and metabolites. The CI spectra were obtained on a Finnigan Model 1015 GC/MS and the library data incorporated into a Finnigan Model 6000 GC/MS interactive data system (5).

These pioneering studies demonstrated the applicability of GC/MS for use in the clinical toxicology laboratory. However, the research instruments used were too sophisticated, expensive, and complicated for the routine clinical toxicology laboratory. The basic research in this area required the simultaneous efforts of many team members at specialized facilities of mass spectrometry with funds from a variety of grants.

In 1979, Gochman et al. reviewed several first-generation specialized GC/MS suitable for use in the clinical laboratory such as the Vitek Olfax 11A and the Hewlett-Packard 5992 (6).

The Vitek Olfax was the first system specifically designed for use in the clinical laboratory. It employed electron impact for the ionization mode, a quadrupole mass analyzer, and a computer containing an 8008 Intel microprocessor, 4K in random access memory (RAM), with an additional 60K available on a tape storage device. The computer controlled the scanning for the mass spectrometer and continuously adjusted critical instrument parameters including the temperature programs for the gas chromatograph. It also provided the user with interactive programs for identification of unknown compounds employing probability-based matching. The Vitek's main limitations were its mass range limit of 400 amu and its slow scanning speed of 150 amu/s. The latter resulted in limited analytical sensitivity. To compensate, the instrument was operated in the selected ion monitoring mode in lieu of total ion monitoring.

The Hewlett-Packard 5992 was the first truly bench-top GC/MS. It had the key feature of inserting the major components of the mass spectrometer into a convoluted diffusion pump, resulting in a lower-cost compact design. With the introduction of a turbo pump and some changes in the computers controlling the instruments and with the introduction of capillary column capability it evolved into the HP 5995 series instruments.

During this decade computer-operated GC/MC instruments have evolved that are easy to operate, relatively inexpensive (less than $50,000), and compact. They are used in a variety of clinical (i.e., hospital or emergency room) and forensic (i.e., employee screening) toxicology laboratories. Two of these mass spectrometers, the Mass Selective Detector (MSD) from Hewlett-Packard and the Ion Trap Detector (ITD) from Finnigan MAT, which is also distributed by Perkin-Elmer, are now widely marketed. The MSD has been extensively used in toxicological studies and it has been evaluated in the literature. The material reviewed in this chapter mainly revolves around these instruments, operated in the electron impact mode. Their ability to separate, using fused-silica capillary gas chromatography, underivatized drugs and metabolites from urine and serum is examined. Also discussed in this chapter are the library search routines available to identify unknown drugs. Their quantitation programs are illustrated and some examples are presented for the confirmation and quantitation of metabolites of some drugs, such as cocaine and tetrahydrocannabinol.

2. MASS SELECTIVE DETECTOR

The major components of the MSD are illustrated schematically in Fig. 4.1. The MSD has an electron impact ionization source producing an

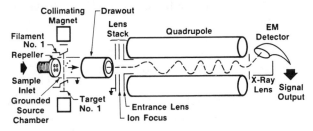

Figure 4.1. The basic components of the mass spectrometer called the *Mass Selective Detector*. Ion source pressure 10^{-4}–10^{-6} torr; electron beam 0.1–1.0 mA, 70 eV electrons; ionization efficiency ~0.01%. (Figure provided by Hewlett-Packard.)

electron energy of 70 eV. Two switch selectable filaments allow continued operation in the event of filament failure. An entrance lens is located downstream of the ion source. The MSD uses a quadrupole mass analyzer which consists of four hyperbolic rods, 203 mm long, each individually machined. The detector consists of a standard electron multiplier that is in the shape of a cornucopia.

The principle of operation of the MSD may be described as follows. The sample enters the ion source via the capillary column where it is bombarded by electrons emanating from the filament. The positively charged ions produced are propelled out of the source and focused into the quadrupole mass filter. When specific DC and AC voltages are applied to the quadrupole rods (in pairs), only ions of a specific "mass-to-charge" ratio will pass from the ion source through the quadrupole to the detector without annihilation. After the ions are separated on the basis of mass-to-charge ratio, they are then counted as they enter the detector. The detected ions can be plotted graphically with the mass on the x axis and the counted ions (abundance) on the y axis to give a mass spectrum. Also, the abundance of all the ions can be plotted as a function of time to give a total ion chromatogram or, alternatively, selected ions can be monitored and plotted as a function of time (selective ion chromatogram). The user may select scan rates from 200 to 1500 amu/s, in 0.1-amu steps. The MSD scans any mass range between 10 and 800 amu.

Two pumps are utilized for maintaining a vacuum in the range of 10^{-6} to 10^{-5} torr: the mechanical rough pump providing the fore-vac and a turbomolecular pump.

For instrument control and data handling the MSD system uses a HP 9000 Series 300 computer, which is referred to as the MS Chemstation. A variety of MSD configurations are available depending upon the options of monochrome or color monitor, disk drive storage capacity, and the type of printer. Other options include the make and model of the gas chromatograph, an automatic liquid sampler (capable of injecting up to 100 liquid samples), and software to permit unattended analyses and customized data reduction.

3. ION TRAP DETECTOR

The ion trap mass spectrometer (ITD) is illustrated schematically in Fig. 4.2. The ion trap consists of a filament assembly, a central cavity defined by ion trap electrodes, and an electron multiplier detector. The filament assembly is composed of two rhenium filaments (one of which is a backup), a lens, and an electron gate assembly. The central compartment

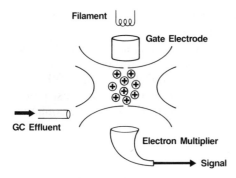

Filament

Gate Electrode

GC Effluent

Electron Multiplier

Signal

Figure 4.2. The basic components of the mass spectrometer called the Ion Trap Detector. (Figure provided by Finnigan MAT.)

consists of three hyperbolic electrodes called the top end cap, the ring electrode, and the bottom end cap. The electron multiplier is the conventional "horn of plenty" design.

The process that occurs in the ion trap can be described as follows. Sample from the capillary GC enters via a heated silica transfer line into a central cavity where the RF voltage is set to storage value. The gate electrode is turned on briefly, allowing electrons into the trap, so that the sample is bombarded by electrons and ionized (electron impact). Recently, new software has been introduced (automatic gain control) that provides automatic variation of ionization time (7). This prevents saturation of the ion trap at high levels of analyte and increases sensitivity at low levels by increasing ionization time up to 25 ms (range from 0.078 to 25 ms). The ions produced by this interaction, which are unique to the sample's molecular structure, are then stored in this central cavity called the ion storage region. An AC voltage is applied to the ring electrode at a constant frequency of 1.1 MHz and variable amplitude (0–7500 V zero-to-peak). Because the frequency of this AC voltage is in the radio-frequency range, it is referred to as the "RF voltage." The ion trap is then scanned by increasing the RF voltage, which causes the positive ions to be ejected sequentially from low to high mass.

The ion trap is described by Finnigan to operate in the mass-selective instability mode. At low RF storage voltage ions have oscillations that are stable; as the RF voltage is ramped, ions of progressively greater mass-to-charge ratio become unstable and are sequentially ejected from the cavity. When the trajectory of ions becomes unstable in the ion trap cavity half are ejected in a collimated beam axially to the electron multiplier, and the rest are ejected at 180° to this direction and neutralized when they strike the grounded top electrode. The full scan range of the ITD is 10–650 amu. The scan rate can be set between 0.25 and 8 scans/

s for the selected mass range. The ion current entering the electron multiplier is multiplied by a factor of about 10^5.

The ITD (filament, mass analyzer, and electron multiplier) is enclosed
in a heated vacuum manifold. It is evacuated by a turbomolecular pump
which discharges into a rotary-vane forepump. Calibration gas (PFTBA)
is admitted into the manifold by a selenoid-operated valve. For instrument
control and data handling the ITD uses an IBM PC-compatible computer
and an Epson compatible printer.

4. IDENTIFICATION AND CONFIRMATION OF UNDERIVATIZED DRUGS

Gas chromatography–mass spectrometry is frequently used in clinical
toxicology (i.e., hospital or emergency room patients) and forensic toxicology (i.e., employee screening programs) for the confirmation or identification of underivatized drugs from extracts of urine or serum. Usually
the GC/MS has been employed for confirmation of drugs detected by other
screening procedures such as immunoassay, thin-layer chromatography,
or gas chromatography. It is also used to identify compounds detected,
but not identified, by other procedures, such as thin-layer chromatography. In some laboratories, where there are not large numbers of samples, GC/MS has been used as the sole method of drug detection (8).

4.1. Urine Blank

Samples are commonly prepared for GC/MS by liquid–liquid or solid
phase extraction techniques. Shown in Fig. 4.3 is an example of a total
ion chromatogram of an acidic and basic urine extract in which the sample
contains no known drugs. In this case 5 ml of urine was solvent-extracted
using the commercially available Toxi-Tube A and Toxi-Tube B and concentrated on the disks provided with the system for application to the
thin-layer plate (Analytical Systems, Laguna Hills, CA 92653). The disks
were washed with 25 μL methanol/ethyl acetate (1:3 by volume) and 1
μL was injected for analysis on the GC/MS. Although the specimen contains no drug other than caffeine, various minor peaks are detected on
the chromatogram corresponding to endogenous substances in the urine
as well as plasticizers (phthalates) from the urine container. In both total
ion chromatograms of Fig. 4.3 the major peak is the standard (mephobarbital) followed by caffeine in the basic extract and phthalates in the
acidic extract. The compounds most commonly detected in "drug-free
urine" are cholesterol, cholestadiene, oleic acid and other lipids, as well
as caffeine, theobromine, and nicotine with its metabolite cotinine (9, 2).

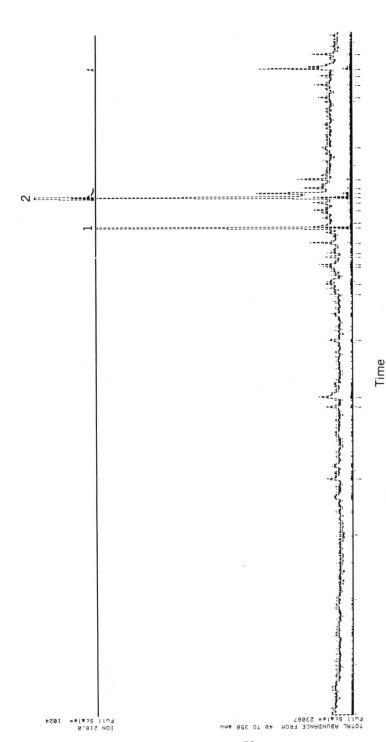

Figure 4.3. Total ion chromatogram of a basic (a) and an acidic (b) extract from urine containing only caffeine. Mephobarbital, the standard, elutes at approximately 10.5 min. In each chromatogram the bottom section was scanned from 40 to 350 amu and the top at 218 amu. Unpublished results from my laboratory. 1 = Caffeine, 2 = mephobarbital; 3 = phthalate.

Time

(a)

ION 218.0
Full Scale= 1024

TOTAL ABUNDANCE FROM 40 TO 350 amu
Full Scale= 23087

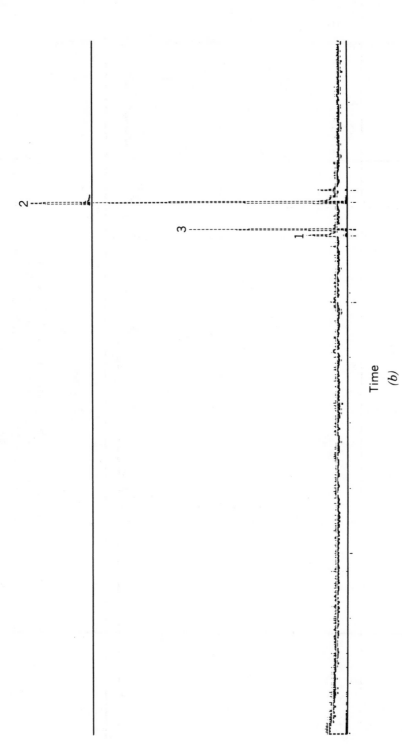

Time

(b)

Figure 4.3. (*Continued*)

4.2.   Detection of Drugs in Urine

Shown in Fig. 4.4 is an example of a total ion chromatogram from a specimen containing compounds frequently found in urine. It was determined, by a computer search of the library and comparison of retention times, that the following substances were present: diphenhydramine and one of its metabolites, acetaminophen, caffeine, salicylate and a metabolite, nicotine, and its metabolite continine. Many of the mass spectra corresponding to minor components are similar to those seen when a urine blank is analyzed (10).

When a medical emergency exists in the clinical toxicology laboratory, it may be necessary to complete an analysis within a few hours or less. Consistent with these needs, GC/MS confirmation procedures that do not require special extractions and derivatization are preferred. Many common drugs fall into this category, for example, codeine, meperidine, propoxyphene, methadone, phencyclidine, acetaminophen, barbiturates, and tricyclics, to name a few. Cocaine also falls in the above category. However, because of its short half-life it is necessary to analyze for one of its longer-lived metabolites. It has become the custom to analyze for benzoylecgonine. Confirmation of benzoylecgonine by GC/MS requires a special extraction and derivatization procedure as described in a subsequent section of this chapter. On the other hand, ecgonine methyl ester, which can also serve as a marker for cocaine use, does not require any special preparation for analysis by GC/MS (11, 12). Ecgonine methyl ester is another major urinary metabolite of cocaine with a slightly shorter half-life than benzoylecgonine (13). Shown in Fig. 4.5 is a total ion chromatogram from our laboratory of a patient who ingested cocaine. This urine sample was extracted with Toxi-Tube A, as described above, and chromatographed on a fused-silica capillary column cross-linked with 5% phenylmethylsilicone. The main components detected were ecgonine methyl ester (EME), caffeine, mephobarbital (the standard), and cocaine.

4.3.   Detection of Drugs in Serum and Plasma

Gas chromatography–mass spectrometry of underivatized drugs in serum and plasma usually is performed to confirm the identity of drugs detected and quantitated by GC. In addition, it can identify unknown or interfering peaks obtained on GC that are contained in its library or alert the operator to the presence of an unidentifiable substance. Methods for GC/MS confirmation of drugs in plasma and serum have been described. Cailleux et al. identified and quantitated neutral and basic underivatized drugs in plasma extracts by GC using nitrogen-specific detectors. They used a GC/

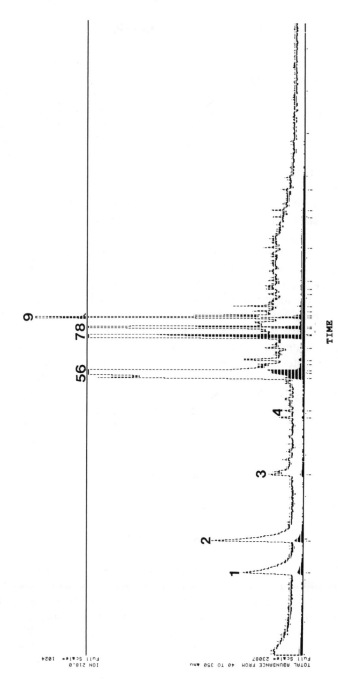

Figure 4.4. Total ion chromatogram of a basic extract from a urine sample containing multiple drugs. Conditions as described in Fig. 4.3. Unpublished results from my laboratory. 1 = Aspirin; 2 = nicotine; 3 = aspirin metabolite; 4 = diphenhydramine metabolite; 5 = cotinine; 6 = acetaminophen; 7 = caffeine; 8 = diphenhydramine; 9 = mephobarbital.

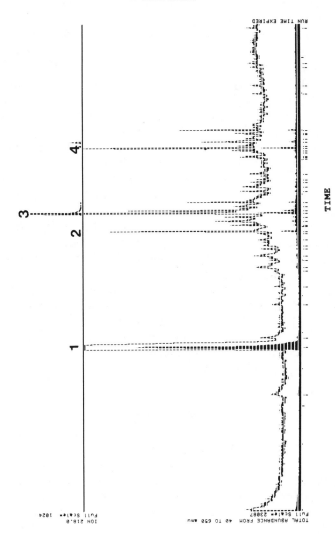

Figure 4.5. Total ion chromatogram of a basic extract from a urine sample containing cocaine and ecgonine methyl ester. Mephobarbital, the standard, eluted at 10.4 min. The bottom chromatogram was scanned from 40 to 650 amu and the top at 218 amu. 1 = Ecgonine methylester; 2 = caffeine; 3 = mephobarbital; 4 = cocaine.

MS equipped with packed glass columns and a chemical ionization source to confirm the identity of drugs detected by GC (14). The complementary use of capillary GC/MS unambiguously to confirm presumptive peaks identified by gas chromatography with flame ionization detection has also been recently demonstrated by Ehresman and Price (15).

In my laboratory we studied various aspects of screening and quantitation of hypnotic-sedatives in serum by capillary gas chromatography with a nitrogen–phosphorus detector (GC/NPD), and confirmation by capillary GC/MS (16). The day-to-day reproducibility of retention times for the extracted mixture of calibration standard hypnotic-sedative drugs analyzed by GC and GC/MS is shown in Table 4.1. The retention times on GC were found to be stable (CV < 1%) and thus useful for identifying these drugs. However, it is known that nonlinearities in retention index may occur at injections of >100 ng (17, 18). In addition, when the number of drugs on the screen was increased to reflect current usage we found some interferences (i.e., coelution). For example, butalbital elutes very closely to butabarbital and chlordiazepoxide has a retention time similar to that of nordiazepam. Concentrations of ethchlorvinyl exceeding 10 μg/ mL interfere with the identification and quantitation of methyprylon. A linear temperature gradient was employed in these studies. A more sophisticated program, similar to the one used by Ehresman and Price (15), would better resolve these components. Although the retention times (Table 4.1) on the GC/MS were found to be less precise than those on the GC/NPD (CV < 3%), this is not critical because the hypnotic-sedatives are mainly identified by their mass spectra in addition to their retention times. Therefore, the identity of a component found on GC is unambiguously established by GC/MS.

5. LIBRARY SEARCH ROUTINES

5.1. Ten Peak

Until recently the library search software for the MSD consisted of a 10-peak search algorithm. All experimental mass spectra were reduced to the 10 most "significant" peaks, as were the spectra in the library. The significance was calculated by multiplying the mass of the peak by its abundance. A routine was then used to calculate the "similarity index" (S.I.) of the unknown spectrum with respect to the library spectra.

In the similarity index calculation, 1.00 represents a perfect match and 0.00 a perfect mismatch. It has been shown that this paradigm sometimes incorrectly identifies the best match or gives very high S.I. values for all

Table 4.1. Day-to-Day Reproducibility of Absolute (ART) and Relative (RRT) Retention Times[a]

Drug and Peak No.	GC Mean ART (min) ± SD	CV (%)	Mean RRT ± SD	CV (%)	GC MS Mean ART (min) ± SD	CV (%)
Methyprylon, 1	7.02 ± 0.03	0.46	0.67 ± 0.010	0.77	8.08 ± 0.24	3.01
Butabarbital, 2	8.20 ± 0.04	0.50	0.78 ± 0.000	0.00	8.95 ± 0.21	2.34
Amobarbital, 3	8.71 ± 0.05	0.52	0.83 ± 0.010	0.62	9.36 ± 0.21	2.24
Pentobarbital, 4	8.95 ± 0.04	0.50	0.85 ± 0.000	0.00	9.50 ± 0.22	2.30
Secobarbital, 5	9.43 ± 0.05	0.49	0.89 ± 0.004	0.54	9.88 ± 0.20	2.02
Glutethimide, 6	10.11 ± 0.05	0.53	0.96 ± 0.000	0.00	10.35 ± 0.16	1.52
Carisoprodol, 7	10.23 ± 0.06	0.55	0.97 ± 0.000	0.00	10.44 ± 0.16	1.54
Mephobarbital, 8	10.54 ± 0.06	0.51	1.00 ± —	—	10.66 ± 0.15	1.43
Phenobarbital, 9	11.06 ± 0.06	0.54	1.05 ± 0.000	0.00	11.12 ± 0.21	1.84
Methaqualone, 10	12.82 ± 0.08	0.59	1.22 ± 0.000	0.00	12.38 ± 0.15	1.24
Diazepam, 11	14.99 ± 0.10	0.64	1.42 ± 0.003	0.27	14.06 ± 0.23	1.66
Nordiazepam, 12	15.52 ± 0.11	0.73	1.47 ± 0.003	0.21	14.28 ± 0.02	0.14

[a] Reprinted with permission of the American Association for Clinical Chemistry from *Clin. Chem. 32*, 35–38 (1986).

Table 4.2. Ten-Peak Library Search

A. For Propoxyphene

Entry	Similarity Index	M_r	Compound
212	0.9973	339.0	Propoxyphene
232	0.9963	149.0	Methamphetamine
373	0.9963	149.0	Methamphetamine
43	0.9959	277.0	Amitriptyline
22	0.9958	339.0	Propoxyphene
352	0.9946	237.0	Ephedrine TMS ether
49	0.9877	255.0	Phenyltoloxamine
178	0.9850	284.0	Psilocybin
173	0.9847	204.0	Psilocin
137	0.9800	328.0	Methotrimeprazine

B. For Methadone[a]

Drug	CAS No.	Library Index No.	Match Quality
1. Methadone	P542 00076-99-3	241	9984
2. Methylephedrine AC	P354	1114	9963
3. Aceprometazine-M AC	P647	1239	9962
4. Mecloxamine	P558 05668-06-4	1078	9950
5. Sotalol	P462 03930-20-9	1368	9933
6. Methylephedrine	P206 00552-79-4	1113	9931
7. Propafenone	P 0 54063-53-5	1692	9928
8. Befunolol	P 0 39552-01-7	1588	9919
9. Aceprometazine	P572 13461-01-3	5	9910
10. Promethazine-M (HO−)	P525	609	9908

[a] Unpublished data from joint experiments conducted at Hewlett-Packard.

the "hits" (10). For example, the result of a library search for propoxyphene is shown in Table 4.2A. Although propoxyphene was correctly identified as the best match by the library search program (0.9973), nine other drug entries had similarity indices ranging from 0.9800 to 0.9963, making it difficult to accept the first choice with confidence. Drugs from such diverse classes as tricyclics, amphetamines, opiate analogs, and hallucinogens were all well matched. A similar situation as that described for propoxyphene occurs when a library is searched for methadone. As shown in Table 4.2B, there are 10 drugs listed with match qualities of better than 99%.

5.2. Probability-based Matching

Recently, Probability-based matching (PBM) library search software has become available for use with the MSD. This software was developed by McLafferty and co-workers over the past 15 years (19–24). A brief description of the main features of the programs and their use with the Chemstation can be found in a handbook published by Hewlett-Packard (25). Some sections from this handbook* are excerpted below:

WHAT IS IT?

This search algorithm compares an unknown spectrum to the reference spectra using a "reverse search" routine. A reverse search verifies that the peaks in a reference spectrum are present in the unknown spectrum. Any extra peaks in the unknown are ignored. This means that a spectrum of a compound mixture may be analyzed.

Since not all mass-to-charge (m/z) values of a mass spectrum are equally likely to occur, the PBM algorithm weighs the importance of both the mass and abundance values to identify the most significant peaks (MSPs) in the reference spectrum. These peaks are then used to create a condensed reference spectrum which is used by the search routine.

The prefilter search routine assigns a significance to each of the peaks in the unknown spectrum and uses these to find the most probable matches in the condensed reference library. The selected condensed spectra are then compared, using the reverse search described above, with the complete unknown spectrum.

U + A VALUES

The PBM search assigns a significance to each mass peak based on both m/z ratio and its abundance. Those peaks with the greatest significance are then used for the matching. The significance of a mass peak is determined by two values, U (uniqueness) and A (abundance).

U (Uniqueness)

Certain masses (m/z) are more likely to occur in a mass spectrum than others. For example m/z = 43 is much more common than m/z = 343. Each mass is assigned a uniqueness value between 0 and 12, the most frequently found masses (m/z = 29, 39, 41, 43) being assigned a value of zero.

*Reprinted by permission of Hewlett-Packard Company from *PBM Search and Parametric Retrieval Software Handbook*, Publication 59973-90003 (1986).

A (Abundance)

The value of A is assigned based on the relative abundance of a mass in the spectrum. The higher the relative abundance, the greater the A value; the assigned values are -2, -1, 1, 2, 3, 4, and 5.

Both the U and A values were developed by McLafferty and his coworkers from statistical studies of 79,650 spectra of 67,128 compounds.

The probability (P) of finding a peak of a particular mass at any abundance level is given by the following equation:

$$P = 0.5^{(U + A)}$$

The U + A calculations are the basis of the PBM search. They are used in the following ways:

- To define peak significance when selecting the 15–26 peaks that are stored in a condensed reference spectrum.
- In the prefiltering algorithm so that only spectra whose most significant peaks are similar to those of the unknown are selected for comparison.
- As one of the factors for evaluating the similarity of reference spectra to an unknown spectrum.

CONDENSED LIBRARY

The PBM search optimizes the mass spectral data base in two ways: each spectrum is condensed when it is added to the library, and the condensed spectra are indexed for fast, easy retrieval.

Spectra are condensed as follows:

1. Select the number of peaks (N) to be stored (range 15 to 26, directly proportional to molecular weight).
2. Calculate the U + A value (significance) of the peaks in the reference spectrum.
3. Select the N most significant peaks for inclusion in the condensed spectrum. The following criteria must be met:

   - The base peak must be included.
   - The molecular ion must be included.
   - Logical losses (e.g., 35, 36 from molecular ion) must be included.
   - Illogical neutral losses and gains (e.g., eight from molecular ion) are reduced.

4. Based on the U + A values, choose a most significant peak (MSP).

This algorithm ensures that both those peaks that are statistically the most significant and those that are logically significant are retained in the condensed spectrum.

The condensed spectra are ordered in the reference library based on their MSPs.

PREFILTER

Use of the prefilter in the PBM search is optional but it reduces the search time dramatically while sacrificing very little accuracy. The prefilter executes the following steps.

1. Assigns U + A values to all peaks in unknown spectrum.
2. Orders the peaks in descending significance (decreasing U + A values).
3. Selects those unknown ions whose U + A values fall within the following range:

    Max U + A > = Ion U + A > = (Max U + A − Delta U + A)

where

Max U + A is the most significant peak of the unknown
Ion U + A is the ion in question
Delta U + A is user-determined (range 0–3)

4. Continues to select ions as in step 3 until either M (user-determined, range 1–999) ions are selected or there are no more ions.
5. For each ion selected, searches the condensed library for all reference spectra that have that one's m/z for an MSP.
6. Does a linear search of the condensed library for those spectra whose MSPs lie outside the scan range (user may choose to skip this step).

COMPARISON ALGORITHM

After the prefiltering is complete, the selected condensed reference spectra are compared with the unknown spectrum. Their similarity is evaluated using the following steps.

1. Normalize the unknown spectrum (all peaks are used).
2. Flag "missing" ions (masses present in the reference but not in the unknown). This does not apply to masses outside the scan range. If there are four or more missing ions, terminate the algorithm and go to next reference spectrum.
3. Calculate RHO values for all peaks in the reference spectrum.

    RHO = unknown abundance/reference abundance

4. Select peak with lowest RHO value.

5. Calculate the confidence factor K using this RHO where K is a sum for all peaks in reference.

$$K = (U + A) + W - D$$

where

U + A = the significance value, precalculated and stored in the reference file
W = the abundance window value, related to percent variation of abundances
D = the dilution factor, proportional to RHO_{min}, the smallest percentage of this compound that could be present in the unknown

6. Calculate the percent contamination, an estimate of purity.
7. Calculate the reliability of the match based on the following.

- K = confidence factors
- dK = $K_{max} - k$, where K_{max} is the result of matching the reference with itself
- N = the number of ions flagged
- Presence or absence of reference molecular ion in unknown

8. Flag the peak just used and get the next reference peak (peak with next lowest RHO). Go back to step 5 for the three next lowest RHO values or until three or more ions have been flagged.
9. If specified, apply tilting functions and go back to step 3.

A list of the 20 best matches is maintained while searching through the prefilter list of reference scans; the list is reported at the end of the comparison.

GETTING THE MOST OUT OF YOUR PBM SEARCH

Overview

In the majority of cases, the PBM search will retrieve a match from the database with a high match quality, and the unknown can be considered identified with a high degree of confidence. However, no search routine, no matter how sophisticated, can provide a conclusive identification 100% of the time. Consider some of the factors affecting the match quality:

- The quality of the spectra in the database;
- The type of instrument used to collect the spectrum of the unknown and the reference;
- The experimental conditions used to collect the spectrum of the unknown and reference;
- Choice of spectrum for background correction;
- Strategy parameters used during the PBM search.

Several of the factors are related to actual collection of the data, and this is the most important aspect of a successful identification by library search: both the reference and the unknown spectrum must be high quality. For instance, if the signal/noise of the mass spectrum is too low, no amount of background correction or changes in PBM strategy parameters will improve the chances of a good match. Also, a spectrum obtained using a GC as an inlet to the mass spectrometer may be much different than that obtained by using a DIP inlet. That is why, for highest confidence in identification, the search should be performed using a user-created library with reference spectra obtained on the same instrument using the same conditions.

A large amount of information is available in the tabular report of a PBM search, as described below.

Examining Search Results

Prob A value representing the probability that the unknown is correctly identified as the reference. This value takes into account all the information below. It is an empirical determination; that is, the K, dK, etc. values are used to "look up" a probability in a table. This "look-up" table was developed by McLafferty and co-workers by extensive evaluation of search results using a large database. Differences in probability values of ±5 are generally not significant. Values less than 50 mean that substantial differences exist between the unknown and reference, and the match should be regarded with suspicion. Values greater than 90 are essentially perfect matches.

K The confidence factor. This value typically ranges from 15–250 and is related to the number of mass peaks in the reference spectrum and the reference's similarity to the unknown. Generally, higher values imply a greater similarity between the unknown and reference.

dK The difference between a "perfect match" (K_{max}, obtained by matching the reference with itself) and the confidence factor for the unknown. A perfect match of the unknown and reference has a dK of 0. Generally, a low dK value is indicative of a "good match," with few differences between the unknown spectrum and the reference. In many instances, a low dK value is of more diagnostic value than a high K value.

Flag The number of ions present in the condensed reference spectrum but not in the unknown. A value of 0 is very good, but a value of 3 does not necessarily indicate a poor match.

Tilt A factor applied to mass spectral abundances to correct for data taken on the sides of chromatographic peaks or for differing responses of mass spectrometers used to obtain the unknown and reference spectra. Tilts of −1, 0, and 1 are usual.

% A measure of the purity of the spectrum as calculated by PBM. PBM is designed to look for components of mixtures. Thus, low purity (small scaling factor) is assumed at first, and then successively higher scaling factors are tried until the best match of reference and unknown is found. A value of 100% means that the base peaks of the reference and unknown were at the same mass.

Con   Contamination factor, yet another measure of purity. This ranges from 0–100, with 0 being the preferred value. Generally, the % and Con factors have opposite behavior, that is, when % is high, Con is low.

C-I   The Class I contamination-corrected reliability value. A value of 100 indicates an exact match. However, since mass spectrometers often cannot distinguish between isomers or members of a homologous series, this value is always less than 100.

R-IV   The class IV reliability value. The match can be an exact match, or an isomer or member of a homologous series, i.e., it is representative of the diagnostic power of a mass spectrometer. The value ranges from 0–99, with high values indicative of a good match.

5.3.   Incos

The ITD uses the Finnigan library search algorithm (Incos™) which incorporates spectrum-condensation techniques and presearch, forward search, and reverse search routines as described in the ITD Operations Manual (26) and in an earlier application report (27).

Some sections from the ITD manual* are excerpted below:

OVERVIEW

Once you have selected the search library and the type of search, the Library Search program performs the following sequence of operations:

1. One scan of data is read from a data file and that mass spectrum is *reduced* to its chemically significant peaks.
2. A *presearch* quickly selects a relatively small set of entries from the library for the main search.
3. A main search is performed, in which each entry selected by the presearch is examined and tested. If it is outside certain limits, it is rejected.

    For each remaining entry, three numberical estimates of its similarity to the unknown are computed.
4. At the end of the search a list of the (up to 10) entries most similar to the unknown is flashed on the screen.

* From *Perkin-Elmer Ion Trap Detector Operating Instruction Manual*, Publication Number 0993-7173. Revision E, 1987. Reprinted by permission of Perkin-Elmer Co.

REDUCTION

The reduction algorithm selects peaks likely to be chemically significant. LIBR (NB). LIB was created by saving 20 peaks per 100 u. The Library Search program's reduction parameter is 40 peaks per 100 u. With this setting, the program will save more peaks from the unknown on each pass than would have been saved if the scan had been entered in LIBR (NB). Then, 16 peaks from the unknown are compared with the biggest eight peaks from each library entry.

The reduction works like this:

1. All peaks have their intensities weighted by their masses and their square root taken. That is, the intensity I of the peak at mass M is replaced by the square root of the dot product of M and I. The weighting by mass gives added importance to the high-mass peaks, which are more unique to the unknown compound and therefore serve to identify it unambiguously. The program calculates the square root of each weighted intensity for the fit algorithm.

2. The 16 largest peaks in the weighted spectrum are saved as likely to have chemical significance.

   Two passes are made through the peaks. The first pass filters out minor peaks and leaves a collection of peak clusters. The second selects the largest peaks in each cluster. Three parameters help to select the number of peaks saved. The first parameter—the number of peaks per 100 u to be saved during the reduction process—is the key one. The second parameter is a window with a half-width of 50 u for the first pass of the reduction; the third parameter is a window with a half-width of 7 u for the second pass.

3. In the first pass, each peak is considered in turn to be the center of a mass window at most 101 u wide. A peak is rejected during the first pass if there are 40 ($= (40/100) \times 101$) larger peaks *in its own window*. (The first peak in the spectrum has a mass window only 50 u wide, so it is rejected if there are 20 ($= (40/100) \times 51$) larger peaks in the 50 u above it. Similar remarks apply to all peaks near the beginning or end of the spectrum.)

4. For the second pass, most of the minor peaks have by now been eliminated, producing, for most spectra, a series of relatively separated clusters. The second pass selects the largest peaks in each cluster. The algorithm of the first pass is repeated, but the mass window is narrowed (± 7 u). Thus, a peak will be rejected during the second pass, if there are six ($= (40/100) \times 15$) larger peaks in its $15 - u$ neighborhood. After the second pass, the 16 largest peaks saved in step 2 are added to the list of peaks, if they are not already present.

Steps 1–4 produce a reduced spectrum that may be used for a library search. The number of peaks per 100 u is set high to ensure that enough peaks will be saved for an accurate search.

PRESEARCH

The presearch quickly selects a relatively small number of entries to be used in the main search. The presearch algorithm compares the 16 most intense peaks of the unknown with the 8 largest peaks in each library spectrum to find at least the minimum number of library spectra selected (the default is 50) that most resemble the unknown. During the presearch, the program accesses the library index file, which contains the 8 largest peaks in any entry.

If the presearch finds 50 (or the minimum selected) or more library spectra with 8 matches, it selects them for further processing. Otherwise, the number of peaks required to match is successively reduced from 8 to 7 to 6, etc., until at least 50 matching entries are found. This information is presented as a table on the display.

The presearch table gives some diagnostic information. If you are searching 75 compounds with 7 or more matches, you are likely to find the unknown or a very similar compound in the library. If you are searching 50 compounds with 2 or 3 matches, you are not.

MAIN SEARCH

Before the similarity calculations take place, the library peak intensities are modified by local normalization. This process corrects for the differences that may exist between two spectra of the same compound acquired in different ways. For example, if a scan from low to high mass is taken on the leading edge of a GC peak, the spectrum will discriminate in favor of high mass ions, because the sample concentration is increasing in the ion trap during the scan. The spectrum of an up-scan made on the trailing edge of the same peak will emphasize low-mass ions.

Local normalization first multiplies the library intensities by a global normalization factor to bring its average intensity near that of the unknown. It then examines each library peak, comparing peaks in a window around each peak in the unknown and the corresponding window in the library spectrum and multiplying peak amplitudes locally to normalize the library spectrum to the unknown. The program does not normalize peaks differing from the corresponding library peaks by more than a factor of 2, and it does not allow their amplitudes to affect the normalization factor. Only peaks at the same nominal masses in both spectra are normalized.

A purity, fit, and Rfit are computed for each library entry. The library entries with the highest purities, fits, or Rfits can be selected and displayed in turn with the <F1> through <F10> keys. Ranking is based on the type of search that you selected. The program will display the spectrum and list the name, entry number, molecular weight, and formula of each of the (up to 10) library spectra that most resemble the unknown, along with fit, Rfit, and purity values for each. Each of the three values is an integer from 0 to 1000.

SELECT SEARCH TYPE

At the heart of the library search program is the algorithm that computes a numerical measure of the similarity of two spectra. For any unknown spectrum and library spectrum, the algorithm produces three numbers—purity, fit, and reverse fit—each with a value between 0 and 1000. The purity search is adequate for most applications. Within doublets or in badly contaminated samples, you may get better results with the fit search. The Rfit search is occasionally useful for identifying components of compounds not found in the library.

PURITY SEARCH

To select a purity search, type **P.**

A *purity* search measures the resemblance of the currently selected data to the specified library entry. A purity of 0 indicates no peaks in common; a purity of 1000 indicates identical mass lists and exactly proportional, locally normalized peak intensities. A purity of 800 suggests that the two compounds have closely related mass spectra. A purity of 600 or more suggests that the two compounds have many fragment ions in common.

FIT SEARCH

To select a fit search, type **F.**

A *fit* search measures the degree to which the library spectrum is included in the unknown spectrum. A fit of 1000 indicates that all library peaks occur as peaks in the unknown and that, for those common peaks, all intensities are exactly proportional. A high fit (800–900) with a lower purity (500–600) suggests that the unknown spectrum is a mixture that includes the compound selected from the library, or that the two compounds have some major substructure in common.

A library search that sorts on fit is called a *mixture search*. A mixture search can provide quite different results from the purity search, which finds the spectra most likely to be identical to the unknown spectra. A mixture search can identify individual components of an unknown mass spectrum that is a mixture of components and is not at all similar to any single entry in the library.

REVERSE FIT SEARCH

To select a reverse fit search, type **R.**

The reverse fit search (*Rfit*) measures the degree to which the unknown spectrum is included in the library spectrum. A fit or Rfit of 800 or more implies a close resemblance between the components. A value of 600 or more implies important shared structural features.

The algorithm that computes purity, fit, and Rfit makes maximum use of the intensity information available for each peak. Perhaps its unique feature is that it provides the number *fit* that allows you to search the library for entries that are likely to be subspectra of the unknown spectrum.

Purity is a measure of the similarity of the unknown to the library spectra. *Fit* is a measure of the degree to which the library spectra are contained in the unknown. *Rfit* is a measure of the degree to which the unknown spectrum is contained in the library spectrum. Purity = fit exactly when all unknown peaks occur in the library entry. Purity = Rfit exactly when all library peaks occur in the unknown.

6. LIBRARIES

Unfortunately, a specialized drug library has not been released during the last few years with the MSD software and one has to cope with the NBS library's IUPAC nomenclature for the more than 42,000 compounds or obtain a library from another source. To solve this problem, Hewlett-Packard has released the library of Pfleger, Maurer, and Weber (28) which contains approximately 1700 drugs, poisons, and their metabolites. The HP Drug Library includes both the S.I. and PBM forms of the data base.

The ITD uses the NBS library as well as a specialized drug library [Finnigan MAT LIBR(TX)] containing approximately 800 compounds with their common names. The library spectra (up to 50 masses for each one) were obtained on a quadrupole mass spectrometer.

Regardless of the library one employs for the toxicology laboratory it must be customized. In some cases, misidentification and/or poor matching of experimental spectra with those in the drug library can be improved by replacing a library entry with a spectrum from a standard acquired on the instrument. Furthermore, of course, drugs not included in the existing library have to be added. For example, the Pfleger library, which was compiled in West Germany, does not have phenylpropanolamine as an entry. Finally, the drug library has to be appended to contain metabolites of drugs. This is particularly important if the GC/MS is going to be utilized for analysis of drugs that occur in the urine.

Until recently, our laboratory utilized a 375 drug library (called Drug Library 1) provided with an early version of HP's software. In order to make this library useful for routine use it was modified by replacing entries with authentic standards, appending it with drugs omitted from the original library, and appending it to contain metabolites of various drugs. Addition of metabolites to the library is both difficult and time consuming. The most direct approach is to acquire a standard of a metabolite from a drug

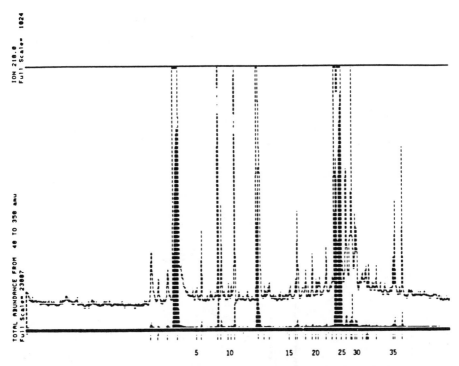

Figure 4.6. The GC/MS total ion chromatogram (40 to 350 amu) from a patient who overdosed on diphenhydramine. The numbering of the spectra on the abscissa was added to the computer printout. Reprinted with permission of the American Association for Clinical Chemistry from *Clin. Chem. 31,* 741–746 (1986).

company. When this is not available one can use a urine sample from a patient overdosed on the drug and/or a literature search. The latter involves searching other published libraries such as the *CRC Handbook of Mass Spectra of Drugs* (29). For a review and discussion of the various collections of mass spectral data see Foltz (30). If the published libraries do not contain the desired spectra of a drug it is usually possible to obtain some information by reviewing original studies performed by drug manufacturers. However, many of these early studies did not always yield mass spectra of well resolved, underivatized metabolic products because capillary columns were not yet available. In practice, as described in the cases below, a combination of approaches may be employed to customize a library.

Shown in Fig. 4.6 is an example of a total ion chromatogram from a patient who ingested 1200 mg of diphenhydramine (Benadryl, Compoz).

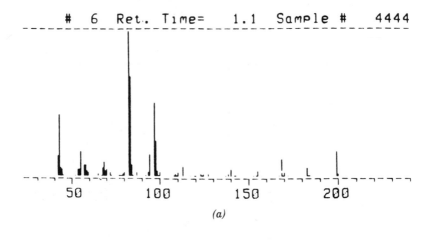

(a)

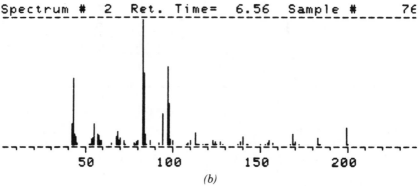

(b)

Figure 4.7. Mass spectrum of ecgonine methyl ester obtained by (a) Ambre et al., 1982. (11) and in my laboratory (b). (a) Reproduced from the *Journal of Analytical Toxicology* by permission of Preston Publications a division of Preston Industries, Inc.

Although the four major chromatographic peaks were identified as di-phenhydramine (spectrum 25) and amphetamine, methamphetamine, and apobarbital (spectra 24, 4, 12), it was shown that the last three were actually metabolites of diphenhydramine. Identification of the metabolites was deduced from a combination of the case study, which involved an overdose of only one drug, and from the literature, which indicated the ions characteristic of the diphenylcarbinol moiety (31).

Another example of using a combination of a case study and literature search to customize the library is shown in Fig. 4.7. In this figure, a mass spectrum obtained in my laboratory from a patient who ingested cocaine (see Fig. 4.5 for the total ion chromatogram) and one published by Ambre et al. for ecgonine methyl ester (11) are compared. The similarity of the two mass spectra allows this metabolite, obtained from a peak resolved on a total ion chromatogram from a basic urine extract, to be added to the library even though no standard was available for injection.

7. PROBABILITY BASED MATCHING VS INCOS

The results of a Hewlett-Packard PBM search using the Pfleger library and the Finnigan MAT Incos search using the Finnigan MAT LIBR(TX) for a 50-ng sample of diphenhydramine are shown in Fig. 4.8. Both searches correctly identify the unknown as the first choice. However, the PBM appears more selective in that it lists only one component with high confidence (any match with less than 15% probability is omitted) whereas the Incos lists many compounds with good matches (1000 is a perfect hit), with the four highest-ranked compounds being diphenhydramine or a closely related substance. These results are typical of those found for other drugs that have been tested on the two systems. A test mixture of nine drugs was fractionated on a Hewlett-Packard and Perkin-Elmer gas chromatograph and a total ion chromatogram (from 40 to 400 amu) obtained on the MSD and ITD, respectively, as shown in Fig. 4.9. In Table 4.3 are shown the results of the PBM search on the Pfleger library and in Table 4.4 are shown the results of the Incos search using the Finnigan MAT LIBR(TX) for this drug mixture. The Incos search ranks 10 compounds for each search and, as shown in Table 4.4, many of the highly ranked compounds give good hits by the criteria of fit or Rfit. Both systems matched the drugs to the correct library entry in all cases (phenobarbital was not detected on the Hewlett-Packard system). As mentioned above, PBM lists fewer possible library entries with good matches for each drug.

With both systems it has been observed that the library search sometimes mismatches compounds to yield an incorrect identification. This was observed with the PBM search for phenylpropanolamine using the NBS library and with the Incos search with a 10-ng injection of amobarbital.

Burns and Rodacy (32) compared combinations of different data base libraries and search algorithms on 48 organic compounds. They found that the Finnigan Incos (on a library of 38,791 mass spectra) and PBM (on a library of 79,560 mass spectra) were about equally successful (approximately 65%) in matching these compounds. In addition they concluded that no organic functional group discrimination was detected in the library search results; there are often markedly different or even in-

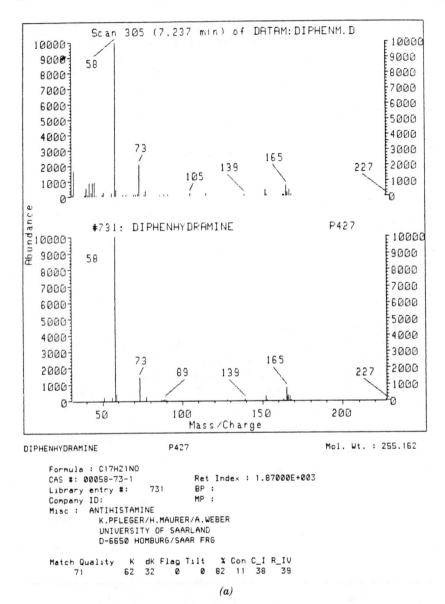

DIPHENHYDRAMINE   P427   Mol. Wt. : 255.162

Formula : C17H21NO
CAS #: 00058-73-1   Ret Index : 1.87000E+003
Library entry #:   731   BP :
Company ID: MP :
Misc : ANTIHISTAMINE
        K.PFLEGER/H.MAURER/A.WEBER
        UNIVERSITY OF SAARLAND
        D-6650 HOMBURG/SAAR FRG

Match Quality   K dK Flag Tilt   % Con C_I R_IV
      71   62 32   0   0   82 11 38   39

(a)

Figure 4.8. Mass spectra of diphenhydramine on Hewlett-Packard and Finnigan instruments with PBM (a) and Incos library (b) search reports.

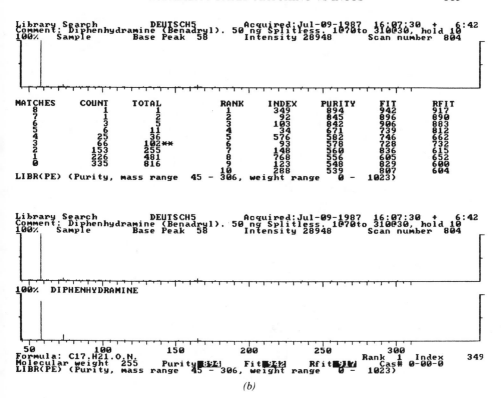

Figure 4.8. (*Continued*)

correct spectra in different libraries; spectral redundancy may be beneficial to help overcome the instrument-dependent m/z abundance intensities; and identification failures are related to both the differences in the library entries and search algorithms.

For positive identification of a drug it is necessary to confirm that it has the correct retention time as well as to establish the similarity of its spectrum to the library selected spectrum. Neither of these systems, per se, has a search paradigm which includes retention in the calculation of match quality. However, a user-contributed program (custom "macro") allows the retention time or retention index of the unknown to be compared with the retention time (retention index) of any drug match found by PBM. With the MSD one may also search libraries for compounds based upon retention time using "parametric retrieval" software. The ITD's automatic quantitation program, a standard feature of the software, allows one to include retention time in conjunction with the library search.

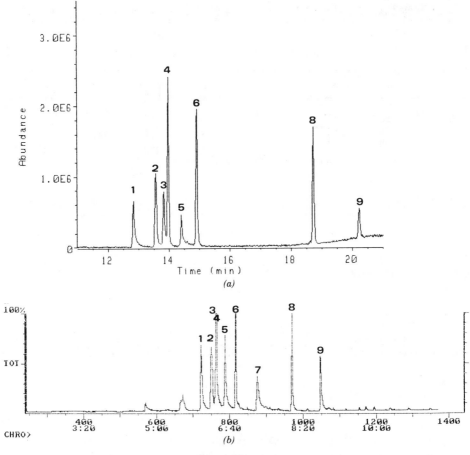

Figure 4.9. Total ion chromatograms of a drug mixture using the MSD (a) and ITD (b) mass spectrometers. 1 = butabarbital; 2 = amobarbital; 3 = pentobarbital; 4 = meperidine; 5 = secobarbital; 6 = glutethimide; 7 = phenobarbital; 8 = cocaine; 9 = codeine. Phenobarbital was not detected using the GC column with the MSD system.

In our laboratory we have a "short list" posted of the retention times and retention indices (calculated relative to mephobarbital as the internal standard) for the drugs most frequently confirmed by GC/MS. These are used in conjunction with the library search for positive identification (Table 4.5).

8. QUANTITATION

With the establishment of federal and state regulations for laboratory accreditation in workplace toxicology, cutoff levels for confirmation of

Table 4.3. PBM Library Search[a]

Name	Prob	CAS No.	Ref. No.	K	dK	Flag	Tilt	%	Con	C-I	R-IV
1. Butabarbital	78	00125-40-6	149	66	52	2	0	90	0	55	14
2. Amobarbital	78	00057-43-2	47	73	52	2	0	100	0	55	18
Pentobarbital A.O.	70	00076-74-4	837	61	53	2	0	100	10	42	12
3. Pentobarbital A.O.	78	00076-74-4	837	82	82	2	-2	100	1	55	19
Butabarbital	25	00125-40-6	149	34	84	0	0	62	50	7	15
Butabarbital-M (HO−)	15		150	26	102	0	0	52	58	3	13
4. Pethidine	83			76	83	1	0	88	30	37	74
5. Secobarbital	83	00076-73-3	961	95	49	2	-3	61	2	57	21
6. Glutethimide	95	00077-21-4	791	101	24	0	0	45	23	53	97
7. Cocaine	73	00050-36-2	465	84	67	1	0	74	41	33	80
8. Codeine	66	00076-57-3	473	90	78	3	-1	100	20	31	45
Hydrocodone	25	00125-29-1	238	33	140	3	0	100	42	8	13

[a] Unpublished data from joint experiments conducted at Hewlett-Packard.

Table 4.4. Incos Library Search[a]

Matches	Count	Total	Rank	Index	Purity	Fit	Rfit
			1. Butabarbital				
8	1	1	1	29	657	941	682
7	4	5	2	135	638	688	696
6	3	8	3	186	615	882	660
5	1	9	4	175	610	922	649
4	16	25	5	28	601	917	617
3	17	42	6	2	568	812	672
2	64	106**	7	122	553	801	673
1	187	293	8	509	432	682	567
0	523	816	9	311	389	598	509
			10	136	357	582	587
			2. Amobarbital				
8	3	3	1	2	642	928	681
7	3	6	2	186	634	921	656
6	4	10	3	135	579	825	628
5	1	11	4	29	571	901	596
4	8	19	5	509	545	846	602
3	14	33	6	28	540	815	559
2	66	99**	7	175	531	853	575
1	187	286	8	122	504	795	573
0	530	816	9	311	413	660	556
			10	796	390	637	597
			3. Pentobarbital				
8	1	1	1	186	789	969	798
7	5	6	2	2	731	907	780
6	4	10	3	175	716	927	742
5	2	12	4	28	701	888	736
4	12	24	5	135	690	870	763
3	14	38	6	29	689	899	739
2	74	112**	7	509	686	887	715
1	154	266	8	122	624	819	738
0	550	816	9	796	545	722	627
			10	311	523	685	595
			4. Meperidine (Demerol)				
8	0	0	1	121	644	791	665
7	2	2	2	812	555	589	649
6	1	3	3	804	391	552	538
5	1	4	4	31	247	334	446

Table 4.4. (*Continued*)

Matches	Count	Total	Rank	Index	Purity	Fit	Rfit
4	5	9	5	182	241	439	458
3	20	29	6	602	223	321	630
2	84	113**	7	537	221	358	425
1	200	313	8	454	218	866	225
0	503	816	9	84	216	302	336
			10	617	209	333	580

5. *Secobarbital (Quinalbarbitone)*

Matches	Count	Total	Rank	Index	Purity	Fit	Rfit
8	2	2	1	127	751	943	781
7	2	4	2	169	706	905	729
6	2	6	3	665	657	847	682
5	2	8	4	30	610	804	638
4	8	16	5	591	595	796	713
3	40	56**	6	798	399	470	737
2	90	146	7	786	334	520	511
1	186	332	8	144	323	473	541
0	484	816	9	480	305	555	347
			10	120	287	494	448

6. *Glutethimide (Doriden)*

Matches	Count	Total	Rank	Index	Purity	Fit	Rfit
8	0	0	1	208	884	931	888
7	3	3	2	118	514	596	589
6	1	4	3	298	454	509	468
5	2	6	4	287	427	499	445
4	23	29	5	190	415	685	528
3	35	64**	6	763	348	502	495
2	69	133	7	733	288	480	387
1	200	333	8	734	278	467	322
0	483	816	9	279	263	653	364
			10	189	262	662	313

7. *Phenobarbital (Luminal)*

Matches	Count	Total	Rank	Index	Purity	Fit	Rfit
8	0	0	1	130	832	946	853
7	0	0	2	799	518	636	778
6	3	3	3	636	443	756	539
5	5	8	4	189	375	543	417
4	18	26	5	381	370	607	386
3	55	81**	6	118	275	682	324
2	112	193	7	287	263	716	321
1	212	405	8	190	239	748	260
0	411	816	9	208	235	613	258
			10	277	231	344	565

Table 4.4. (*Continued*)

Matches	Count	Total	Rank	Index	Purity	Fit	Rfit
			8. Cocaine				
8	1	1	1	172	904	963	917
7	1	2	2	801	571	645	656
6	2	4	3	652	448	677	613
5	5	9	4	386	346	668	396
4	15	24	5	582	345	878	384
3	28	52**	6	163	335	801	349
2	88	140	7	156	330	560	386
1	206	346	8	391	282	475	560
0	470	816	9	209	264	656	321
			10	224	254	562	289
			9. Codeine				
8	0	0	1	72	753	903	805
7	1	1	2	802	553	700	681
6	0	1	3	124	515	793	605
5	3	4	4	580	348	688	463
4	0	4	5	477	267	457	387
3	14	18	6	558	235	443	264
2	53	71**	7	228	224	400	284
1	224	295	8	178	206	660	252
0	521	816	9	596	179	295	313
			10	606	165	334	436

[a] Previously unpublished data from joint experiments conducted at Perkin-Elmer.

some drugs in urine have been established as well as minimal levels for spiking of proficiency test specimens. Quantitative GC/MS confirmation is being instituted as a way to gauge the performance of laboratories on proficiency tests, although it is recognized that the results probably do not have any physiological significance in samples of urine. Shown in Table 4.6, for example, are the concentrations proposed by the National Institute of Drug Abuse for federal testing (33, 34). The MSD comes with convenient packages for quantitation of drugs using the selective ion monitoring (SIM) mode of operation. The user has a choice of report formats when using the internal standard method, including automatically determining the ratios of selected analyte ions ("qualifiers") as a check for interfering substances. An example of a report containing this information is shown in Fig. 4.10. The dimethyl derivative of 11-nor-delta-9-tetrahydrocannabinol-9-carboxylic acid (THC-COOH) and of trideuterated

Table 4.5. Retention Times and Library Locations on GC/MS

Drug[a]	Absolute Retention Time	Relative Retention Time	GC/MS Library Location Numbers
Acetaminophen	9.18	0.850	183, 115
Acetaminophen metabolite	9.40	0.816	183, 115
Amitriptyline	12.48	1.180	43
Amphetamine	4.32	0.392	35, 186
Amobarbital	9.22	0.871	378, 185, 25
Aprobarbital			169, 184
Butabarbital	8.82	0.834	387
Butalbital	8.86	0.837	216, 213, 159, 67
Caffeine	10.30	0.948	340, 59
Carisoprodol (2)	7.86	0.743	376, 65
Carisoprodol (2)	10.36	0.979	376, 65
Chlorodiazepoxide	14.32	1.354	360, 200, 23
Cocaine	13.00	1.182	142, 191
Cocaine metabolites (3)	7.14	0.641	391
(Ecgonine methyl ester)	7.50	0.673	391
(Ecgonine methyl ester)	7.86	0.716	391
Codeine	13.80	1.282	192, 42
Cotinine	9.16	0.848	394
Desipramine	12.78	1.208	44
Dextromethorphan	12.46	0.966	392
Diazepam	13.92	1.316	361, 24, 218
Diphenhydramine	10.58	0.980	86, 63, 383, 384,
Diphenhydramine metabolites			383, 384, 385
Doxepin (2)	12.56	1.187	388
Doxepin (2)	12.68	1.198	388
Doxylamine	11.00	1.009	393
Glutethimide	10.24	0.968	197, 28
Imipramine	12.64	1.195	66
Meperidine	9.72	0.900	114, 247, 195
Metamphetamine			373
Methadone	12.40	1.146	139, 202
Methaqualone	12.28	1.161	386
Methprylon	7.92	0.747	9
Meprobarbital (IS)	10.70	1.00	167, 316
Meprobamate (2)	7.14	0.675	62, 201
Meprobamate (2)	9.80	0.926	62, 201
Nicotine	5.10	0.562	278, 56
Nordiazepam	14.26	1.350	375
Nortriptyline	12.58	1.189	47
Normeperdine	9.96	0.920	248
Pentazine	12.92	1.221	235

Table 4.5. (*Continued*)

Drug[a]	Absolute Retention Time	Relative Retention Time	GC/MS Library Location Numbers
Pentobarbital	9.40	0.888	206, 27
Phencyclidine	10.56	0.994	208, 140
Phenobarbital	10.98	1.038	158, 209
Phenylpropanolamine	6.02	0.570	377
Phenytoin	13.62	1.258	164, 196
Propoxyphene	12.40	1.172	212, 22
Propoxyphene metabolite	10.62	1.004	379. 380, 381, 382
Secobarbital	9.76	0.922	214, 144

[a] Numerals in parenthesis indicate number of metabolites.

THC-COOH (the internal standard) were chromatographed and the ions for the metabolite (313, 357, and 372 amu) and internal standard (316, 360, and 375 amu) counted. When quantitation is performed using peak qualifiers, a calculated result is reported only if the confirming ions are found in the proper retention time and if their ratios, relative to the quantitating ions, fall within an expected range (35). Dozens of publications have shown the utility of the MSD's quantitation programs for drugs of abuse.

The ITD also comes with convenient software for internal standard

Table 4.6. Threshold Levels for Proficiency Testing and GC/MS Confirmation

Drug Class	Specimen Composition	Concentration (ng/mL)	
		Proficiency Test Samples	GC/MS Cutoff
Marijuana	THC metabolite (Δ^9-THC-9-carboxylic acid)	18	15
	Benzoylecgonine	180	150
Opiates	Morphine	360	300
	Codeine	360	300
Amphetamines	Amphetamine	600	500
	Methamphetamine	600	500
Phencyclidine	Phencyclidine	30	25

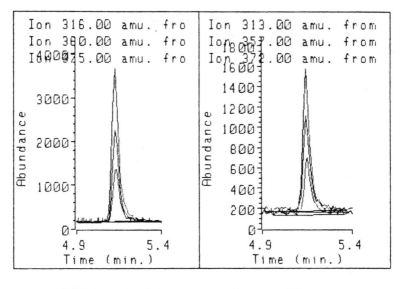

	D3-THC internal standard			THC	
Ions	Area Ratio	Expected Range	Ions	Ratio	Expected Range
360/316	0.57	0.54 - 0.80	357/313	0.74	0.54 - 0.80
375/316	0.36	0.28 - 0.42	372/313	0.43	0.28 - 0.42

Ion	Retention Time (min)	Expected Range	Ion	Retention Time (min)	Expected Range
316	5.15	5.07 - 5.27	313	5.16	5.06 - 5.27
360	5.15	5.07 - 5.27	357	5.16	5.06 - 5.27
375	5.16	5.07 - 5.27	372	5.17	5.06 - 5.27

Figure 4.10. A Hewlett-Packard report format for selective ion monitoring of dimethyl derivative of 9-THC-COOH and trideuterated 9-THC-COOH.

method quantitation and customizing reports. It performs quantitation in either the full scan mode or the multiple peak monitoring mode. The full scan mode has the advantage of gathering mass spectral data to improve certainty of identification. The practical application of this approach depends on the ITD giving improved sensitivity over conventional mass spectrometers in the total ion current mode. The capability of the ITD's quantitation programs has not been evaluated in the literature, although it has been used to measure ion ratios for the detection of 9-THC-COOH (36).

When SIM is used, it has been pointed out by Foltz (37), many factors must be considered regarding what constitutes a positive identification and accurate quantitation. These factors include the derivatization and extraction procedures, the chromatography, the ionization process (which is restricted to electron ionization at present with the instruments described here), the number of ions monitored, and the uniqueness of the mass spectra.

As internal standards for GC/MS quantitation, stable isotope analogs, when available, are generally favored over unlabeled homologs or isomers. Although studies have not been performed for specific drugs of abuse comparing deuterium-labeled analogs to unlabeled homologs, there is substantial evidence that the former makes assays more precise, accurate, and specific, although a report to the contrary has been published (38). Although stable isotope analogs may also increase recoveries by serving as a carrier, they may lead to imprecision if the amount used is too large relative to the analyte (39).

9. DERIVATIZED DRUGS

Many compounds are more effectively analyzed by GC or GC/MS if they are first converted to a derivative. Some of the GC/MS procedures for the more common drugs and metabolites in clinical and workplace toxicology have been reviewed (30, 40, 41). One may also consult, for example, *The Journal of Analytical Toxicology,* which publishes many papers dealing with the development and evaluation of GC/MS procedures for drug confirmation. In this journal, for example, Mulé and Casella recently published GC/MS confirmation procedures for cocaine, morphine, codeine, amphetamine, methamphetamine, and phencyclidine (42). The U.S. military has conducted urine drug screening for many years. The methods used by the Navy, developed for the common drugs-of-abuse, are available in their Standard Operating Procedure Manual (SOPM). In addition, many manufacturers who produce solid phase sorbents for extraction frequently provide detailed procedures for extraction and derivatization (e.g., Bond-Elut, Analytichem Int.; Prep cartridges, DuPont Company; Narc columns, J.T. Baker; Fisher PrepSep, Fisher Scientific; drug extraction columns from Biochemical Diagnostics, Applied Separations, Worldwide Monitoring Corporation). Some selected methods for electron-impact GC/MS are indicated below.

9.1. Delta 9-THC-COOH

One of the most popular procedures for derivatization of 9-THC-COOH involves treatment with iodomethane and tetramethylammonium hydroxide (TMAH), as the catalyst (43, 44). The dimethyl derivative of 9-THC-

COOH is monitored at 313, 357, and 372 amu whereas the trideuterated internal standard is usually monitored at 316 or 375 amu for quantitation (see Fig. 4.10). The 9-THC-COOH glucuronide conjugate must be hydrolyzed before derivatization.

Another procedure for 9-THC-COOH by GC/MS uses an adaptation of a method described by Foltz (40) employing liquid–liquid (45) or solid phase extraction (46). The derivatizing agent is *N, O*-bis(trimethylsilyl)trifluoroacetamide + 1% trimethylchlorosilane (BSTFA + TMCS), which yields the trimethylsilyl (TMS) derivative. This compound is monitored with the mass spectrometer at 371, 473, and 488 amu for 9-THC-COOH and at 374 or 491 amu for the trideuterated internal standard. A variety of other procedures are available for 9-THC-COOH in urine and some of these have recently been reviewed (40, 41, 37, 47).

9.2. Benzoylecgonine

The most common method for GC/MS analysis of benzoylecgonine involves esterification producing the *N*-methyl (48) (cocaine), *N*-ethyl, or *N*-propyl derivative (40). These can be formed by heating in the appropriate alcohol at acid pH, by heating with *N, N*-dimethylformamide di-*N*-alkylacetal, or, for the ethyl derivative, by using TMAH and ethyl iodide at room temperature. For quantitation and confirmation, deuterated benzoylecgonine is available as an internal standard from Merck Sharp & Dohme Isotopes or from the Research Triangle Institute. *M/e* values for GC/MS monitoring of propylbenzoylecgonine are 210, 272, and 331 amu and 213, 275, and 334 amu for the deuterated internal standard. More recent methodology utilizes solid phase separation procedures and silylating reagents for derivatization (e.g., see Worldwide Monitoring Corporation).

9.3. Opiates

One of the most common derivatization methods for the identification and quantitation of codeine and morphine by electron-impact GC/MS employs acetylation (49, 50). After hydrolysis of the glucuronide conjugate, a liquid–liquid extraction is performed and acetic anhydride is employed for acetylation. Nalorphine is used as an internal standard. The ion pair ratios 341 and 282 of acetylcodeine, 369 and 327 of diacetylmorphine, and 395 and 353 of diacetylnalorphine are monitored and the ratios for an unknown must give a response within prescribed limits for positive identification. As discussed by Paul et al. (49), many other methods for derivatization of morphine and codeine including silylation, pentafluorobenzylation, tri-

fluoroacetylation, pentafluoropropionylation, and heptafluorobutyryla-
tion have been described.

9.4. Other Drugs

In addition to the journals and reference works cited to above, some GC/
MS confirmation procedures may be found in a technical newsletter called
Toxi-Lab News published by Analytical Systems (Irvine, CA 92718). For
example, one recent issue presented a procedure for sympathomimetic
amines (phenylpropanolamines, pseudoephedrine, amphetamine, and
methamphetamine) after derivatization with pentafluoropropionic anhydride. Another issue of *Toxi-Lab News* presented a procedure for six
benzodiazepines after hydrolysis to their corresponding benzophenones.
Recently, Analytical Systems and Hewlett-Packard jointly published a
monograph entitled *Screening and Confirmation of Drugs of Abuse with
Toxi-Lab TLC and Hewlett-Packard GC/MS Systems.*

With the routine use of GC/MS instruments in more and more clinical
and forensic laboratories, one may expect to see a variety of procedures
developed, for various drugs, that are easier and faster methods than those
employed for research studies.

ACKNOWLEDGMENTS

I gratefully acknowledge the experimental assistance and discussions from the
staff at Hewlett-Packard (Mr. Doug Postl, Mr. Ernie Bonelli, and Drs. Wayne
Duncan, Leslie Hodges, and Patrick Perkins), Finnigan MAT (Dr. Clay Camp-
bell), and Perkin-Elmer (Dr. Adam Patkin).

REFERENCES

1. J. R. Althaus, K. Biemann, J. Biller, P. F. Donaghue, D. A. Evans, H.-J.
   Forster, H. S. Hertz, C. E. Hignite, R. C. Murphy, G. Preti, and V. Reinhold,
   Experientia, 26, 714 (1970).
2. C. E. Costello, H. S. Hertz, T. Sakai, and K. Biemann, *Clin. Chem., 20,* 255
   (1974).
3. N. C. Law, V. Aandahl, H. M. Fales, and G. W. A. Milne, *Clin. Chim. Acta,
   32,* 221 (1971).
4. B. S. Finkle, D. M. Taylor, and E. J. Bonelli, *J. Chromatogr. Sci., 10,* 312
   (1972).

5. B. S. Finkle, R. L. Foltz, and D. M. Taylor, *J. Chromatogr. Sci., 12*, 304 (1974).

6. N. Gochman, L. J. Bowie, and D. N. Bailey, *Anal. Chem., 51*, 525A (1979).

7. R. A. Yost, W. McClennen, and H. L. C. Meuzelaar, Applications Report, Finnigan Mat, 1986.

8. Dr. Bill Clausen, personal communication.

9. P. A. Ullucci, R. Cadoret, P. D. Stasiowski, and H. F. Martin, *J. Anal. Toxicol., 2*, 33 (1978).

10. D. G. Deutsch and R. J. Bergert, *Clin. Chem., 31*, 741 (1985).

11. J. J. Ambre, T. I. Ruo, G. L. Smith, D. Backes, and C. M. Smith, *J. Anal. Toxicol., 6*, 26 (1982).

12. J. Ambre, M. Fischman, and T. I. Ruo, *J. Anal. Toxicol., 8*, 23 (1984).

13. J. Ambre, *J. Anal. Toxicol., 9*, 241 (1985).

14. A. Cailleux, A. Turcant, A. Premel-Cabic, and P. Allain, *J. Chromatogr. Sci. 19*, 163 (1981).

15. D. J. Ehresman and S. M. Price, *J. Anal. Toxicol., 9*, 55 (1985).

16. V. Soo, R. J. Bergert, and D. G. Deutsch, *Clin. Chem., 32*, 325 (1086).

17. L. L. Plotczyk, *J. Chromatogr., 240*, 349 (1982)

18. M. Brogurz, J. Wijsbeck, J. P. Franke, and R. A. Zeum, *J. Anal. Toxicol., 7*, 188 (1983).

19. F. W. McLafferty, R. H. Hertel, and R. D. Villwock, *Org. Mass Spectrom., 9*, 690 (1974).

20. G. M. Pesyna, F. W. McLafferty, R. Venkataraghavan, and H. E. Dayringer, *Anal. Chem., 47*, 1161 (1975).

21. G. M. Pensyna, R. Venkatarghavan, H. E. Dayringer, and F. W. McLafferty, *Anal. Chem., 48*, 1362 (1976).

22. B. L. Atwater, More Reliable Identifications of Unknown Mass Spectra Using the Probability Based Matching Algorithm, Ph.D. Thesis, Cornell University, Ithaca, NY, 1980.

23. K. Mun, D. R. Bartholomew, D. B. Stauffer, and F. W. McLafferty, *Anal. Chem., 53*, 1938 (1981).

24. D. B. Stauffer, Improved Identification of Unknown Mass Spectra Using the Probability Based Matching Algorithm, Ph.D Thesis, Cornell University, Ithaca, NY, 1984.

25. *PBM Search and Parametric Retrieval Software Handbook,* Publication No. 59973-90003, Hewlett-Packard Company, 1986.

26. Perkin-Elmer Ion Trap Detector Operating Instruction Manual. Publication Number 0993-7173, Revision E, 1987.

27. S. Sokolow, J. Karnofsky, and P. Gustafson, Finnigan MAT Application Report Number 2, *The Finnigan Library Search Program,* 1978.

28. K. Pfleger, H. Maurer, and A. Weber, *Mass Spectral and GC Data of Drugs, Poisons and Their Metabolites,* VCH Publishers, New York, 1985.

29. I. Sunshine, *CRC Handbook of Mass Spectra,* CRC Press, Boca Raton, FL, 1981.

30. R. L. Foltz, "Mass Spectrometry," in I. Sunshine, Ed., *CRC Handbook of Mass Spectra of Drugs,* CRC Press, Boca Raton, FL, 1981.

31. T. Chang, R. A. Okerholm, and A. J. Glazko, *Res. Commun. Chem. Pathol. Pharmacol., 9,* 391 (1974).

32. F. B. Burns and P. J. Rodacy, 35th ASMS Conference on Mass Spectrometry and Allied Topics, 389 (1987).

33. "Draft Standards For Accreditation of Laboratories Engaged in Urine Drug Testing," National Institute of Drug Abuse, Revision of April 1987.

34. Alcohol, Drug Abuse, and Mental Health Administration, "Mandatory Guidelines for Federal Workplace Drug Testing Programs; Final Guidelines; Notice,"*Federal Register, 53,* 11970 (1988).

35. Hewlett-Packard, Publication Number 23-5953-8069, 1987.

36. H. H. McCurdy, L. J. Lewellen, L. S. Callahan, and P. S. Childs, *J. Anal. Toxicol., 10,* 175 (1986).

37. R. L. Foltz, "Analysis of Cannabinoids in Physiological Specimens by Gas Chromatography/Mass Spectrometry," in R. C. Baselt, Ed., *Advances in Analytical Toxicology,* Vol. 1, Biomedical Publications, Foster City, CA, 1984, pp. 125–157.

38. M. G. Lee and B. J. Millard, *Biomed. Mass. Spectrom., 2,* 78 (1975).

39. W. A. Garland and M. P. Barbalas, *J. Clin. Pharmacol., 26,* 412 (1986).

40. F. L. Foltz, A. F. Fentiman, and R. B. Foltz, *GC/MS Assays for Abused Drugs in Body Fluids,* NIDA Research Monograph 32, DHHS Publication No (ADM) 80-1014, Superintendent of Documents, Washington, DC (1980).

41. *Drug Analysis Using GC/MS,* Publication No. 23-5955-5391, Hewlett-Packard, 1987.

42. S. J. Mulé and G. A. Casella, *J. Anal. Toxicol., 12,* 102 (1988).

43. J. D. Whiting and W. W. Manders, *Avait. Space Environ., 54,* 1031 (1983).

44. M. L. Abercrombie and J. S. Jewell, *J. Anal. Toxicol., 10,* 178 (1986).

45. T. S. Baker, J. V. Harry, J. W. Russell, and R. L. Myers, *J. Anal. Toxicol., 8,* 255 (1984).

46. R. E. Kiser, H. E. Ramsden, J. M. Patterson, and D. Law, *Clin. Chem., 32,* 1115 (1986).

47. R. L. Hawks and C. N. Chiang, *Urine Testing for Drugs of Abuse,* NIDA Research Monograph *73,* DHHS Publication No. (ADM) 87–1481 (1986).

48. J. E. Wallace, H. E. Hamilton, D. E. King, D. J. Bason, H. A. Schwertner, and S. C. Harris, *Anal. Chem., 48,* 34 (1976).

49. B. D. Paul, L. D. Mell, Jr., J. M. Mitchell, J. Irving, and A. J. Novak, *J. Anal. Toxicol., 9,* 222 (1985).

50. R. E. Strumpler, *J. Anal. Toxicol., 11,* 97 (1987).

CHAPTER

5

SCREENING WITH HIGH PERFORMANCE LIQUID CHROMATOGRAPHY

DENNIS W. HILL AND KAREN J. LANGNER

University of Connecticut
Storrs, Connecticut

1. INTRODUCTION

1.1. Purpose of Screening Procedures

The term "drug screening" cannot be elucidated by a single definition. There are a variety of ways to approach the concept of drug screening and these ultimately depend on the type of information being sought as well as the nature and number of samples being "screened." For example, screening a sample for a broad range of drugs may be quite different from determining whether a specific compound or group of compounds is present in a sample. Regardless of its purpose, the screening procedure should be rapid and determine the presence or absence of the drugs of interest. The efficiency and specificity of the tests chosen to gain this information are interdependent. When large numbers of samples are screened for a broad range of compounds, tests with less specificity must

129

be used in order to maintain timely processing. On the other hand, speed becomes less important when techniques are utilized that determine the presence of one drug or a particular class of drugs to the exclusion of all others.

1.2.   Data Required for Compound Identification

When biological samples are screened for drugs, it is necessary to differentiate between the naturally occurring constituents of the matrix and the exogenous substances of interest. Separation techniques, usually chromatographic, accomplish this phase of the screening procedure. Once separation has been accomplished, it is possible to determine if there is a constituent in the sample that is not endogenous. Separation and detection therefore can be considered the primary part of the screening procedure. The secondary phase involves the identification of an exogenous constituent. Identification of a suspect compound involves the collection of different types of chemical data reflecting different physicochemical properties of the suspect substance. Chromatographic analyses provide one type of data that aids in compound identification, that is, retention characteristics. This information alone, however, is not sufficient to identify a substance. Additional chromatographic analyses may not yield new or unique information about the compound because most retention information obtained from the same substance tends to be related (1). Therefore, it is necessary to perform spectrometric analyses, such as mass spectrometry or ultraviolet spectrometry, which provide structural information. Only when all the data collected uniquely characterizes a compound (drug) from all the other possibilities, is it safe to assign an identity to it. Traditionally, this decision has been made by the individual based on experience and expertise in data interpretation. Recently, however, a mathematical theory has been proposed based on the concept of the number of distinguishable regions (DR) of a particular analytical system (2). Each analytical method produces a definite number of DR (Table 5.1) based on the uncertainty of the system to distinguish, separate, or distribute compounds without overlap or interference. For example, chromatographic methods contain fewer DR than do spectrometric analyses because there inherently are more areas of definition (range) in spectrometric systems. The theory states that "a minimum necessary mathematical condition for identification of a compound is that the number of distinguishable regions must be greater than or equal to the relevant population of compounds in order for absolute identification to be possible."

For example, from Table 5.1, TLC × GC × MS = 18,000 DR. Using

Table 5.1. Distinguishable Regions of Several Analytical Methods

Method	Range[a]	± Uncertainty[b]	Regions
Thin-layer chromatography	5 cm	0.5 cm	5
Gas chromatography	3.2 min	0.16 min	10
Liquid chromatography	24 mL	0.4 mL	30
Ultraviolet spectroscopy	202 nm	2 nm	50
Infrared spectroscopy	3350 cm^{-1}	10 cm	168
Mass spectroscopy	360 amu/esu	0.5 amu/esu	360

[a] Range = expanse of the analytical system.
[b] Uncertainty = expected deviation inherent within the system.

HPLC with UV detection in place of GC in the above analytical sequence results in 2,700,000 DR, which is a significant increase. If the number of DR produced by the combined analyses is larger than the population of compounds, the chemist may assign an identity to the unknown compound with a reasonable degree of confidence.

1.3. HPLC as a Screening Method

In primary screening of large numbers of samples for a broad range of drugs, chromatographic techniques can provide the balance needed between specificity and efficiency. Very specific analytical techniques, such as colorimetric spot tests and immunoassays, may be the method of choice when the population of compounds of interest is known and relatively few; however, they are inappropriate for primary screening of many samples when large numbers of drugs are targeted.

Thin-layer chromatography and GC are specifically suited to many analytical needs, but each has limitations in application. HPLC has been used as a drug screening method for some time and its application for this purpose continues to expand (3–17). It is a highly versatile technique that provides very good resolution and, unlike GC, is not adversely affected by the analyte's polarity or lack of volatility. The use of solvent programming in HPLC makes it possible to screen for a large polar range of compounds in a single system. HPLC can be easily interfaced with visible and UV detectors. The combination of HPLC with UV detection offers a good balance between efficiency and specificity. Being able to obtain both retention and spectral information about a compound within the same analytical process is also an efficient way to gain information about a compound's identity. The use of retention indices further enhances the usefulness of HPLC as a screening procedure.

2. HPLC RETENTION DATA

2.1. Methods of Determining Retention Characteristics in HPLC

HPLC retention data can be expressed as retention time (R_t), capacity factor (k'), relative retention time (RR_t), and retention indices. Each of these data types is affected to varying degrees by differences in column size and composition as well as other variables, such as temperature, composition of mobile phase, and flow rate.

Because it is difficult to duplicate all these parameters precisely in an HPLC system, it is not uncommon to observe different retention data for the same compound. Historically, lack of reproducibility of HPLC data has been a disadvantage of its use in qualitative analysis (18–22). Consistency of chromatographic conditions improves data reproducibility; however, variations in R_t and k' occur with time with a given column using the same mobile phase. This is due to changes in the stationary phase composition. If, however, retention data are normalized to one or more reference compounds, reproducibility should improve.

Separation of drugs by HPLC reverse phase chromatography using solvent programming allows for the identification of a wide range of chemically different compounds in a single system. However, solvent programming can further alter the reproducibility of retention data owing to fluctuations in the rate of change of the solvent strengths during the program. These changes may occur between instruments or for the same instrument over a period of time. Using relative retention data with solvent programming can improve data reproducibility.

The theory of relative retention time and retention index is based on the assumption that changes in HPLC parameters will affect the elution of test compounds and reference compounds to the same degree. Despite variations in the absolute retention volume or retention time of a test compound, the retention of that compound relative to a reference compound should remain constant.

2.2. HPLC Retention Indices

Experience has shown that the above assumption holds true when the reference compound is structurally similar to the test compound and elutes relatively close to it. However, in drug screening with HPLC, it is not possible to find an individual compound that can meet these criteria. Use of a homologous series of compounds as references can solve the problem of finding a close eluting reference compound. These series of compounds vary from each other by the number of a constant functional

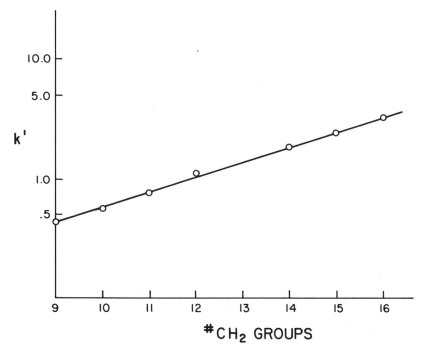

Figure 5.1. Alkylphenone homologous series. Isocratic reverse phase HPLC. Number of functional groups (CH_2) vs. log capacity factor (k').

group, such as a methylene group. Each compound in the series contains one more of the functional group than the preceding compound. In reverse phase chromatography, compounds in a homologous series elute in order of the increase in the number of methylene groups.

Under isocratic conditions, a linear relationship is seen between the number of functional groups in the reference compound and the log of their respective capacity factors (Fig. 5.1). In a solvent program system, a curvilinear relationship exists between the number of functional groups in the reference compounds and their respective retention times (Fig. 5.2). Equations (1) and (2) are used to calculate retention indices for these HPLC systems.

$$\mathrm{RI}_{(i)} = \frac{\log k'_x - \log k'_N \; \Delta N \times 100}{\log k'_{N+1} - \log k'_N} + N \times 100 \qquad (1)$$

where $\mathrm{RI}_{(i)}$ = retention index for compounds eluting in an isocratic HPLC system

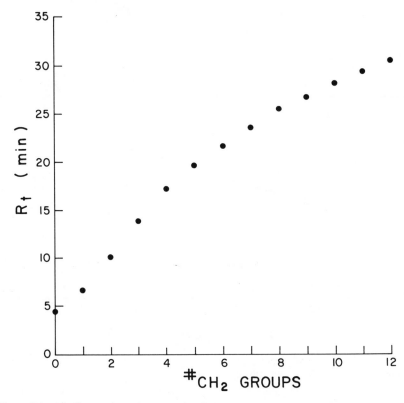

Figure 5.2. Alkylketone homologous series. Solvent gradient reverse phase HPLC. Number of functional groups (CH_2) vs. R_t.

x = test compound

k'_x = capacity factor of test compound

k'_N = capacity factor of reference compound eluting before test compound

k'_{N+1} = capacity factor of reference compound eluting after test compound

N = number of functional groups in the reference compound

ΔN = difference between the number of functional groups in the reference compound eluting before and after the test compound

$$\mathrm{RI}_{(p)} = \frac{t_x - t_N}{t_{N+1} - t_N} \Delta N \times 100 + N \times 100 \tag{2}$$

where $\mathrm{RI}_{(p)}$ = retention index for compounds eluting in a solvent program HPLC system

x = test compound

t_x = retention time of test compound

t_N = retention time of reference compound eluting before test compound

t_{N+1} = retention time of reference compound eluting after test compound

N = number of functional groups in the reference compound

ΔN = difference between the number of functional groups in the reference compound eluting before and after the test compound

In each equation the index number is calculated relative to two of the compounds in the series, the ones eluting just before and just after the test compound. When a homologous series of reference compounds elutes over the whole range of solvent strengths within the system, two of these reference compounds will elute relatively close to the test compound.

Two homologous series of compounds that have been used for HPLC retention indices are the alkyl-2-ketones and the alkylphenones. Each has advantages and disadvantages as a reference series. The alkyl-2-ketones have a greater range of polarities and thus elute over a greater range of mobile phase solvent concentrations than the alkylphenones. The alkylphenones have a much stronger UV-absorbing chromophore than the alkyl-2-ketones and therefore can be detected more easily by UV spectrometry. The use of diode array UV detectors with the alkyl-2-ketones as a reference series diminishes the problem of their low UV absorption. As can be seen in Fig. 5.3a, a chromatogram monitored at 230 nm poorly resolves the alkyl-2-ketones from the base line. However, when the chromatogram is reconstructed at the wavelength of maximum absorbance for alkyl-2-ketones (274 nm) the entire homologous series is easily observed (Fig. 5.3b). This allows HPLC analysis and optimization of wavelength for maximum detection of test compounds as well as reference series.

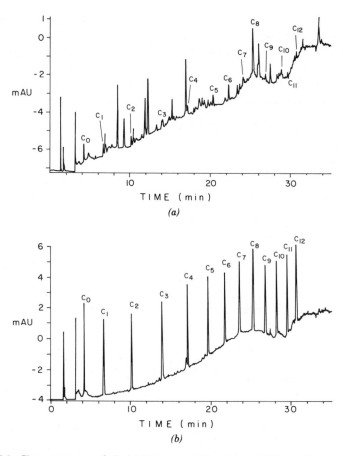

Figure 5.3. Chromatograms of alkyl-2-ketones. (*a*) Monitored at 230 nm; (*b*) reconstructed at 274 nm.

2.3. Problems and Uses of Retention Indices in HPLC

Using retention indices for compound identification is not without some restrictions, however. The assumption that all HPLC parameters will affect both test and reference compounds similarly may not be true in practical applications. Although many of the physical differences in LC systems, such as column length, temperature, and flow rate, may be compensated for by retention indices, variations in the chemical interactions between the stationary phase, mobile phase, and solute may be expected to affect test compounds and reference compounds differently

(23–25). Retention indices can be altered by the use of different column types and packing materials. Solvent concentration changes will also affect retention indices, as will differences in eluent pH. The changes in retention indices observed in these data may be attributed to mixed function separation mechanism. In reverse phase chromatography, using silica-based bonded phases, a mixture of alkyl and silanol functional groups exists. The alkyl groups contribute to the desired solvaphobic mechanism of retention in reverse phase HPLC. However, the silanol functional groups are partially ionized and interact with the ionized solute molecules to form hydrogen bonds, thereby producing an ionic interactive mechanism. This mixed function mechanism generates non-Gaussian-shaped peaks and increases band broadening. The retention of a compound in this system will be a result of the ratio between the degree of interaction of these different mechanisms on the compound in the system. The relative distribution of the two mechanisms will be a function of the ratio of the silanol functional groups to the alkyl functional groups and will vary for different compounds. Organic compounds with ionizable and polar functional groups would be expected to vary more in retention than hydrophobic organic compounds with no polar functional groups. In practice, it is difficult to produce silica-based stationary phases with silanol functions or ratios of silanol to alkyl functional groups that are consistent from column to column. This variability is more pronounced between phases that are alkylated by different procedures, as is the case between column materials produced by different manufacturers. Therefore, the elution of compounds relative to each other would be expected to vary between different columns.

Table 5.2 represents data collected by Smith et al. (26) which compare retention indices for some local anesthetics and test compounds under various HPLC conditions. Small changes in temperature, pH, and solvent concentration resulted in retention indices that were relatively consistent, as indicated by % CV. Comparison of retention index and capacity factor data for the same compounds on different columns (Table 5.3) shows that, even though retention indices vary (% CV range = 2.5–19.6), their reproducibility is much better than that of capacity factor data (% CV range = 43.6–91.8).

Retention indices relative to the alkylphenones for mycotoxins separated on a solvent program HPLC system have been reported by Hill et al. (27). The retention index for this group of compounds was determined on four different Zorbax C_8 columns and showed reproducibility of better than 1% CV (Table 5.4). A method was described whereby the elution of two reference phenones, relative to the other compounds in the homologous series, was calculated and used to determine the retention index

Table 5.2. Effect of Column Temperature, Methanol Concentration and Eluent pH on the Retention Indices (RI) of Local Anesthetics and Column Test Compounds

Compound	Temp.[a]		pH[b]		MeOH Concn[c]	
	Av. RI	% CV	Av. RI	% CV	Av. RI	% CV
Chloroprocaine	465	4.1	469	2.7	467	5.8
Lignocaine	562	1.5	537	1.8	531	0.4
Cocaine	632	1.2	638	1.3	635	2.3
Amylocaine	714	0.8	721	1.1	718	1.5
Benzocaine	809	0.9	794	5.9	809	0.9
2-Phenylethanol	766	0.9	767	0.3	766	0.5
Nitrobenzene	793	0.8	794	0.1	793	1.1
p-Cresol	797	0.1	793	0.4	793	0.6
Toluene	918	1.4	920	0.1	920	2.1

[a] Average RI for analyses at column temperature of 10, 30, and 40°C.
[b] RI for solvent pH of 2.0, 2.5, and 3.0.
[c] Average RI for solvent methanol concentration of 10%, 15%, and 20%.
Reference series—alkylphenones.
Column—Hypersil ODS.
Eluent—methanol:H_2O:1.0% (v/v aqueous orthophosphoric/n-hexylamine.

Table 5.3. Effect of Different Brands of ODS Columns on Capacity Factor (k') and Retention Index (RI) on Local Anesthetics and Column Test Compounds

Compound	k'		RI	
	Av.[a]	% CV ($n = 3$)	Av.[a]	% CV ($n = 3$)
Chloroprocaine	1.76	60.8	534	19.6
Lignocaine	1.76	43.6	542	11.3
Cocaine	7.70	64.8	689	13.3
Amylocaine	10.84	62.9	720	6.8
Benzocaine	28.50	81.1	799	2.5
2-Phenylethanol	9.00	91.8	760	6.8
Nitrobenzene	22.78	80.9	773	2.6
p-Cresol	15.61	67.5	735	7.4
Toluene	84.68	88.4	884	4.0

[a] Average of retention data for analysis on Hypersil, Partisil, and Zorbax ODS columns.
Reference series—alkylphenones.
Eluent—methanol:H_2O:1.0% (v/v) aqueous orthophosphoric/n-hexylamine.

Table 5.4. Retention Index (RI) for Mycotoxins Separated on Four Different Zorbax C$_8$ Columns by Solvent Program HPLC

Mycotoxin	RI	
	Av.	% CV
Aflatoxin G$_2$	764	0.9
Aflatoxin B$_2$	786	0.2
Aflatoxin G$_1$	792	1.0
Parisiticol	796	1.0
Aflatoxin B$_1$	826	0.7
Aflatoxicol I	841	0.7
Dechlorogriseofulvin	873	0.6
o-Methylsterigmatocystin	921	0.6
Roridin A	937	0.6
Verrucarin A	958	0.5
Tetrahydrodeoxyaflatoxin B$_1$	1037	0.4
Paxilline	1228	0.8

Reference series—alkylphenones.
Solvent program mobile phase—20% CH_3CN/H_2O to 100% CH_3CN.
Flow rate—2.0 mL/min for 60 min.

of the compounds of interest. The advantage of this technique was that only two reference compounds had to be co-analyzed with a sample of unknown composition, thus reducing the possibility of co-elution of one of the reference compounds with a sample component.

It has been stated that an advantage of using solvent programming in HPLC analyses is that a broader polar range of compounds can be analyzed in a single system. However, there are circumstances in which it is advantageous to utilize isocratic systems, for instance, for the analysis of a few closely eluting compounds. Under these conditions, it would seem reasonable to assume that a retention index established for these same compounds with solvent programming could be used to predict a retention index under isocratic conditions. There has been good correlation between retention data generated by temperature programming and isothermal conditions in GC, so one might expect similar correlations in HPLC. This premise would also seem to be substantiated if the data reported by Smith et al. in Table 5.2 are considered. Changes in the organic concentration of the mobile phase over a range of 30% produced retention indices that correlated relatively well. Data generated in our laboratory, however, seem to indicate that retention indices produced by isocratic

Table 5.5. Comparison of Retention Indices (RI) for Mycotoxins Separated by Isocratic (I) and Gradient (G) Solvent Systems

Compound	RI (I)	RI (G)
Aflatoxin G_2	637	764
Aflatoxin B_2	680	786
Aflatoxin B_1	718	826
Dechlorogriseofulvin	772	873
Roridin A	818	921
o-Methylsterigmatocystin	832	937
Verrucarin A	862	958
Tetrahydrodeoxyaflatoxin B_1	1010	1037

Reference series—alkylphenones.
Column—Zorbax C_8.
Mobile phases—Isocratic system: 50% CH_3CN/H_2O
Gradient system: 20% CH_3CN/H_2O to 100% CH_3CN in 60 min.

systems are markedly different from those produced by solvent gradient systems (Table 5.5).

The difference in variability observed by Smith from the data presented in Table 5.5 may be attributed to the presence of n-hexylamine in the system used by Smith. The organic primary amine would have a tendency to associate with the free silanol groups in the stationary phase, thus reducing their ability to interact with the polar solutes eluting from the LC system. Primarily, the single solvaphobic mechanism would be responsible for the separation of compounds, and it might be expected that this would favor a constant relative retention between compounds of various polarities.

3.   HPLC/UV SPECTRAL ANALYSIS

3.1.   UV Photodiode Array Detection with HPLC

Spectral data, as shown in Table 5.1, reveal more specific information about a compound than chromatographic data. For compound identification to be successful, one or more spectrometric analyses are needed. Mass spectra and infrared spectra generally yield more information than UV spectra; however, because HPLC systems are more easily interfaced with UV detectors, the combination of HPLC and UV spectrometry is a highly desirable screening method. The recent introduction of photodiode array UV spectrometers has removed some of the difficulties previously encountered with the comparison of UV spectral data. The addition of computerized library search routines to compare spectral data further enhances the specificity and efficiency of screening by HPLC/UV.

One of the problems that arises in using HPLC as a general screening method for drugs is determining an optimal wavelength to monitor the eluent that has high sensitivity for all the compounds that may be of interest. Photodiode array detectors can monitor absorbance at many wavelengths simultaneously. This provides the analyst with the ability to reconstruct chromatograms at any or all of the wavelengths over a chosen wavelength range. This is particularly advantageous for analyses in which several or many compounds may be present with differing maximum absorbances, as is often the case with biological samples.

Figure 5.3 illustrates this concept. The top chromatogram was constructed at a wavelength of minimum absorbance for a reference series of alkylphenones. The bottom chromatogram was reconstructed at the optimal wavelength for these compounds, where they are easily detected. Graphic display of the contour map of a chromatogram will also allow visualization of eluted compounds with very low absorbance at the monitored wavelength. Examining a spectrum in this way allows the operator to see areas of minimal absorbance and reconstruct the chromatogram at a wavelength where that particular compound is more easily detected.

Another difficulty in general chromatographic screening procedures is the possibility of co-elution of two or more compounds. Peak purity can be determined by examining spectral profiles collected at several different places along the peak, usually one on the upslope, one at the apex, and one on the downslope. Because the ratio of concentrations of different compounds is expected to be different during elution, a visual overlay of these spectra would show deviations in the spectra if more than one compound eluted within the peak.

Another way of determining peak purity is to plot the ratio of absorbance at two wavelengths against the elution time over the area of the peak. If only one compound has eluted within the peak area, this ratio should be constant. Fluctuations in the ratio indicate that more than one compound has eluted.

3.2. Computerized Spectral Comparison for Compound Identification

Computerized library search routines to compare spectra have been used in mass spectrometry and infrared spectrometry for some time (28–35). Recently this approach has been successfully adapted for use with UV spectra (36–37). The UV library search routine reported by Hill et al. (37) consists of two phases: a presearch system, which narrows the possible number of matches of unknown spectra to library spectra to a few, and a point-by-point comparison of the test spectrum to the presearch choices existing in the spectral library. The presearch algorithm is based on a comparison of wavelengths of maximum absorption and their FTA (fraction of total absorbances) values. The absorbance values at each wavelength in the spectrum are normalized to the total area under the spectral curve by the following equation:

$$N_i = A_i \left/ \left(\sum_{j=200}^{402} A_j \right) \right.$$

where N is the FTA at wavelength i, A is the absorbance at wavelength i, i is the individual wavelength, and j is the wavelengths in the spectral profile. Spectra with FTA values at wavelength maximum values that are close to those of the test spectra are chosen from the library. These presearch choices are then compared to the unknown spectrum at each point in the curve by means of the following equation, which is based on the scaled sum of the differences between the unknown and library spectra:

$$M = \frac{2 - \sum |S_i - R_i|}{2} \times 1000$$

where M is the goodness of fit value (FIT), S is the response in the sample spectra at wavelength i, and R is the response in reference spectra at wavelength i.

The efficiency of this program to choose the correct spectra from a library of more than 350 drugs has been very good. Even compounds with similar structures such as procaine and benzocaine are differentiated by this search routine. In 31 out of 34 samples, the correct compound (procaine) was chosen over benzocaine and placed as second-best FIT in the remaining three sample analyses.

Figures 5.4 and 5.5 are examples of actual laboratory samples analyzed using HPLC for separation, a UV photodiode array spectrometer for detection, and the computer library search routine for identification. The analysis in Fig. 5.4 is a spiked proficiency test sample for urine toxicology submitted by the College of American Pathologists (CAP). Figure 5.5 is based on data collected from a human urine sample. Both analyses were performed on extracts obtained directly from a general screening paired-

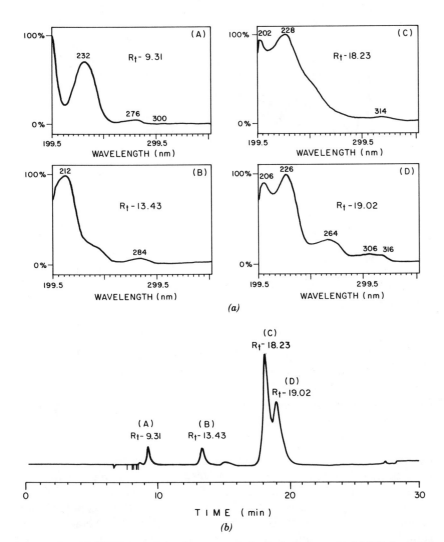

Figure 5.4. HPLC/UV spectral data from extract of spiked urine sample. (*a*) UV absorption spectra for the four peaks; (*b*) chromatogram showing order of elution. HPLC Parameters: Column—Hamilton PRP-1 (250 mm × 4.6 mm), Solvent A—1% NH₄OH, Solvent B—1% NH₄OH/CH₃CN, Solvent Program—0–100% B/A 30 min, Flow rate—2.0 mL/min.

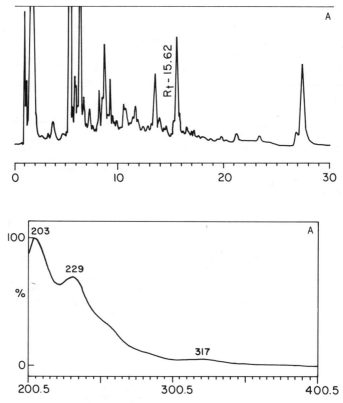

Figure 5.5. Chromatograms (top) and UV spectra (bottom) of (A) human urine extract and (B) oxazepam standard. HPLC Parameters same as for Fig. 5.4.

ion separation procedure used routinely in our laboratory (38). Both figures compare chromatograms and UV spectra of samples to those of reference standards. The computer search routine indicated best FITs for all the correct compounds in the CAP test sample. The computer search results for these compounds are shown in Table 5.6. In the actual urine sample, the correct compound (oxazepam) was chosen as the second-best FIT. The retention time of oxazepam is closer to the retention time of the sample peak and further analyses by TLC and mass spectrometry confirmed the identity of this compound to be oxazepam.

4.   CONCLUSION

Over the past few years HPLC has grown rapidly in popularity as an analytical method. This is probably due to its versatility. As a screening

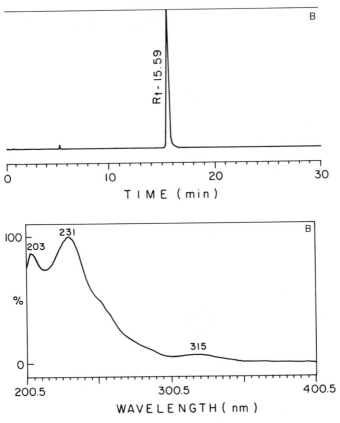

Figure 5.5. (*Continued*)

tool, it has advantages over TLC and GC. Although TLC is a cost- and time-efficient method for screening large numbers of samples, it does not provide very high resolution and is limited in the range of compounds that can be analyzed with a single solvent system.

Gas chromatography provides the highest resolution of compounds and allows volatile compounds to be separated easily and quickly. Groups of compounds can be analyzed in the same system if temperature programming is used. Gas chromatographs are easily interfaced with a variety of detectors. The major disadvantage of using gas chromatography as a primary screening method is that nonvolatile and highly polar compounds are not suited to GC without additional adjustments to the sample. This method therefore does not provide the best balance between efficiency and specificity at the initial screening process, but can be highly useful for analysis of specific types of compounds.

Colorimetric assays and immunoassays can be very effective screening

**Table 5.6. Computer Search Routine Results for UV
Spectral Data from Extract of Spiked Urine Sample
(See Fig. 5.4)**

Rec. No.	Fit	Drug
	Spectrum A	
212	937	Benzoylecgonine
92	788	Acetylsalicylic acid
326	777	Quinidine
388	777	Quinine
	Spectrum B	
280	970	Ethylmorphine
273	962	Codeine
275	873	Didrate
	Spectrum C	
410	993	Nordiazepam
276	980	Diazepam
291	970	Flurazepam
298	948	Oxazepam
	Spectrum D	
270	955	Methaqualone
230	878	2-Hydroxyquinoline
43	838	Chlorpropamide

methods if a single compound or a group of structurally related compounds is being targeted. They are therefore restricted in their application.
Immunoassays can also pose additional difficulties. There is a danger of
cross-reactivity occurring between the antibody and extraneous compounds in the sample, resulting in false positives. Another problem that
may occur with immunoassays is the susceptibility of the antibodies to
denaturation, which reduces or eliminates their reactivity with the specified compound. Factors that could result in denaturation of the antibodies
are high ionic strengths and extreme pH. Intentional addition of salts,
acids, or bases to the biological fluids being assayed can produce false
negative results.

HPLC can be adopted for use in separating many structurally different
compounds. Commercial columns are readily available and on-line collection of spectral data from HPLC systems is easy to accomplish. The

combination of HPLC separation and diode array UV spectrometers is a highly useful analytical tool for qualitative analysis of drugs and toxic compounds in biological samples. Technology continues to improve both the sensitivity and efficiency of the instruments available for this type of analysis.

Increasing demands for rapid, accurate drug screening procedures have led to the rapid development and improvement of these HPLC techniques. Research is continuing in the area of computerization of library searching for drug identification and the use of retention indices as a means of standardizing HPLC data. Future improvement of these techniques can provide data that will be comparable among different laboratories, thereby further enhancing the accuracy, efficiency, and application of HPLC as a drug screening method.

REFERENCES

1. K. E. Elgar, *The Identification of Pesticides at Residue Concentrations,* Advances in Chemistry Series 104, The American Chemical Society, Washington, DC, 1971.
2. W. Duer, J. de Kanel, and T. D. Hall, Proceedings of the Third International Symposium on Equine Medication Control, Lexington, KY, June 12–14, 1979, pp. 486–496.
3. P. J. Cashman, J. I. Thornton, and D. L. Shelman, *J. Chromatogr. Sci., 11,* 7 (1973).
4. J. E. Evans, *Anal. Chem, 45,* 2428 (1973).
5. J. H. Knox and J. Jurand, *J. Chromatogr., 82,* 398 (1973).
6. I. Jane, *J. Chromatogr., 111,* 227 (1975).
7. P. J. Twitchett, *J. Chromatogr., 104,* 205 (1975).
8. S. Kitazawa and T. Komura, *Clin. Chem. Acta, 73,* 31 (1976).
9. R. F. Adams and F. L. Vandemark, *Clin. Chem., 22,* 25 (1976).
10. P. M. Kabra, B. E. Stafford, and L. J. Morton, *Clin. Chem., 23,* 1284 (1977).
11. P. M. Kabra, H. Y. Koo, and L. J. Morton, *Clin. Chem., 24,* 657 (1978).
12. T. Daldrup, F. Susanto, and P. Michalke, *Fresenius Z. Anal. Chem. 308,* 413 (1981).
13. B. Law, R. Gill, and A. C. Moffat, *J. Chromatogr., 301,* 165 (1984).
14. R. Gill, R. W. Abbott, and A. C. Moffat, *J. Chromatogr., 301,* 155 (1984).
15. R. J. Flanagan and I. Jane, *J. Chromatogr., 323,* 173 (1985).
16. I. Jane, A. McKinnon, and R. J. Flanagan, *J. Chromatogr., 323,* 191 (1985).
17. R. Erge, R. K. Muller, and H. J. Wehran, *Die Pharamazie, 40,* 153 (1985).
18. John K. Baker, *Anal. Chem., 51,* 1693 (1979).

19. J. K. Baker, R. E. Skelton, T. N. Riley, and J. R. Bagley, *J. Chromatogr.*, *18*, 153 (1980).

20. R. M. Smith, T. G. Hurdley, R. Gill, and A. C. Moffat, *Chromatographia*, *19*, 401 (1984).

21. R. M. Smith, T. G. Hurdley, R. Gill, and A. C. Moffat, *J. Chromatographia*, *19*, 407 (1984).

22. R. M. Smith, T. G. Hurdley, R. Gill, and A. C. Moffat, *J. Chromatogr.*, *355* (1986).

23. H. Magg and K. Ballschmiter, *J. Chromatogr.*, *331*, 245 (1985).

24. J. K. Baker, *J. Liq. Chromatogr.*, *4*(2), 271 (1981).

25. J. K. Baker and C. Ma, *J. Chromatogr.*, *169*, 107 (1979).

26. R. M. Smith, *J. Chromatogr.*, *236*, 313 (1982).

27. D. W. Hill, T. R. Kelley, K. J. Langner, and K. W. Miller, *Anal. Chem.*, *56*, 2576 (1984).

28. H. S. Hertz, R. A. Hites, and K. Bieman, *Anal. Chem.*, *43*, 681 (1971).

29. S. R. Lowry and D. A. Huppler, *Anal. Chem.*, *55*, 1288 (1983).

30. M. F. Delaney and P. C. Uden, *Anal. Chem.*, *51*, 1242 (1979).

31. A. Hanna, J. C. Marshall, and T. L. Isenhour, *J. Chromatogr. Sci.*, *17*, 434 (1979).

32. L. V. Azarraga, R. R. Williams, and J. A. de Haseth, *Appl. Spectrosc.*, *35*, 466 (1981).

33. J. A. de Haseth and L. V. Azarraga, *Anal. Chem.*, *53*, 2292 (1981).

34. G. Hangac, R. C. Wiebolt, R. B. Lam, and T. L. Isenhour, *Appl. Spectrosc.*, *36*, 40 (1982).

35. M. D. Erickson, *Appl. Spectrosc.*, *35*, 181 (1981).

36. A. F. Fell, B. J. Clark, and H. P. Scott, *J. Chromatogr.*, *316*, 423 (1984).

37. D. W. Hill, T. R. Kelley, and K. J. Langner, *Anal. Chem.*, *59*, 350 (1987).

38. D. W. Hill, T. R. Kelley, S. W. Matiuck, K. J. Langner, and D. E. Phillips, *Anal. Lett.*, *15*(B2), 193 (1982).

CHAPTER

6

MICROCOLUMN AND DIRECT-SAMPLE-ANALYSIS LIQUID CHROMATOGRAPHY

STEVEN H. Y. WONG

University of Connecticut School of Medicine
Farmington, Connecticut

Therapeutic drug monitoring (TDM) and toxicology have become well established specialties in the clinical laboratory (1). Until recently, the predominant method was the immunoassay, complemented by liquid, gas–liquid, and thin-layer chromatography, as well as other techniques. Because of much recent publicity regarding illicit drug abuse, there is a resurgence of interest in using chromatography as a screening and/or confirmation technique. In fact, a tentative proposal of guidelines from the National Institute of Drug Abuse (2) recommends that for laboratories engaging in substance abuse testing, definitve confirmation should be performed by gas chromatography/mass spectrometry (GC/MS).

Although GC/MS is the preferred definitive method, other chromatographic techniques in combination with selective detection may fulfill the requirements. Liquid chromatography (LC), owing to the lack of a universal detector, has thus far been underutilized as a drug screening tool. However, for TDM and toxicology drug analysis, LC offers the following advantages: specificity, sensitivity, and flexibility (3). In order to focus on these advantages, this chapter deals with two current areas, microcolumn and direct-sample-analysis LC for drug analysis. These methods are particularly suited for limited-sample-size applications such as neonatal and pediatric drug monitoring. Although neither of these techniques is used routinely, the clinical application of microcolumn LC, microbore (MBLC) in particular, has been recently explored (4, 5), and various approaches of direct-sample analysis such as column switching have been attempted for some time (6). If the latter methods are successfully applied in combination with computer control and data reduction capability, their ease and reliability might match those of automated immunoanalyzers. And they may be manipulated to offer turnaround times of an hour or less. Thus by reviewing these techniques, we shall be better able to define their potential merit in relation to the immunoassay and other techniques.

Two volumes (7, 8) and a recent review article (4) have addressed MBLC and its applications up to 1985. Recently, I described the possible clinical application of MBLC for the analysis of theophylline, caffeine, procainamide, and N-acetylprocainamide (5) as well as its use for the assay of chloramphenicol (9).

In the area of direct-sample analysis, there is renewed interest. This technique may be performed by the following approaches: microinjection of serum, typically less than 20 µL, onto columns with various pore size and functionalities (10, 11); "traditional" column switching (6); micellar chromatography, recently pioneered by DeLuccia et al. (12); bimodal, internal surface reverse phase LC, pioneered by Pinkerton (13); and liquid chromatography/electrochemical detection (LCEC) with optional postcolumn photolytic derivatization as proposed by Krull (14).

In view of the controversial debate on the best and most practical way to perform drug screening, and the potential need of TDM and toxicology analysis for neonatal and pediatric patients, this chapter attempts to review, with emphasis on the chemistry of analysis, the following areas of MBLC: (1) a brief introduction to the theory; (2) a survey of current instrumentation and a review of literature with special consideration of potential clinical applications of MBLC; (3) instrumentation; and (4) MBLC assays of theophylline, caffeine, procainamide, and N-acetylprocainamide, as well as chloramphenicol with reference to its clinical pharmacology. For direct-sample analysis, the following areas are reviewed: (1) recent examples of microinjection; (2) column switching and on-line column extraction (extraction tube and solvent extraction); (3) micellar chromatography; (4) electrochemical detection with photolytic derivatization; and (5) bimodal, internal surface reverse phase.

1. MICROCOLUMN ANALYSIS

1.1. Theoretical Considerations

According to Scott (15), 1976 marked a renewal of interest in microbore HPLC. Prior to that, Horvath, Preiss, and Lipsky (16) explored the use of small-bore columns packed with pellicular particles. The advantages of microbore HPLC include low solvent consumption and enhanced mass sensitivity. Table 6.1 from Scott's chapter shows the relative cost of using columns with different column internal diameters (ID) ranging from 0.51 mm to the standard 4.6-mm ID columns. A cost saving of 20-fold is achieved when using the 1.02-mm instead of the 4.6-mm column.

The reason that microbore HPLC provides enhanced mass sensitivity

Table 6.1. Solvent Consumption for Columns of Different Diameters

Column Diameters (ID)	Flow Rate for a Linear Velocity of 0.14 cm/sec (mL/min)[a]	Volume Used in an 8-hr Day (mL)	Volume Used in 250-Day Year (L)	Cost/Year/ Chromatograph (Distilled-in-Glass Heptane at $28/gal) ($)
0.51 mm (0.020 in.) 1/16 in. OD	0.012	6.9	1.5	15
0.76 mm (0.30 in.) 1/16 in. OD	0.027	13	3.3	21
1.02 mm (0.040 in.) 1/16 in. OD	0.044	24	6.1	39
1.29 mm (0.050 in.) 1/16 in. OD	0.079	38	9.5	60
1.59 mm (0.063 in.) 1/8 in. OD	0.12	57	14	89
4.6 mm (0.181 in.) 1/4 in. OD	1	480	121	769

[a] 0.14 cm/sec = flow rate of 1 mL/min through a column 25 cm long and having a dead volume of 3 mL. (Reproduced with permission from Ref. 15.)

is derived from the following equations for estimating the minimum detectable mass (m):

$$m = \frac{V_\mathrm{p}C}{2} \tag{1}$$

where C = minimum detectable concentration

V_p = peak width of solute at the peak base

Further derivation would show that:

$$m = \frac{2\pi r^2 l\psi(1 + k')C}{\sqrt{N}} \tag{2}$$

where r = column radius

l = column length

ψ = fraction of mobile phase-occupied volume

k' = capacity factor of the solute

N = column efficiency in theoretical plates

Equation (2) shows that m depends directly on the square of the radius of the column. According to Table 6.2 from Scott (15), the relative mass sensitivity of the 4.6-mm and 1.0-mm ID columns would be 232 and 11 ng, respectively—a 20-fold increase in sensitivity. In order to minimize extracolumn dispersion owing to sample valve, connecting tubes, and

Table 6.2. Solvent Consumption, Mass Sensitivity, and Minimum Apparatus Dispersion Necessary for Columns of Different Diameters

Column ID (mm)	Column Dead Volume (mL)	Relative Solvent Consumption	Relative Mass Sensitivity (ng)	Column Variance (μL^2)	Apparatus Variance (μL^2)	Apparatus Standard Deviation (μL)
4.6	11.6	1	232	13533	1353	36.8
2.0	2.2	0.19	44	483	48.4	7.0
1.0	0.55	0.047	11	30.2	3.3	1.7
0.5	0.14	0.012	2.8	1.9	0.2	0.4

[a] Column length 1 m, efficiency 40,000 theoretical plates, K' of solute = 1, and detector sensitivity = 10^{-6} g/mL. (Reproduced with permission from Ref. 15.)

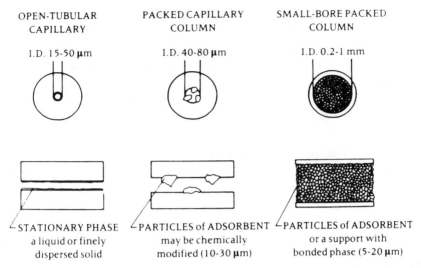

Figure 6.1. Types of microcolumns used in HPLC. (Reproduced with permission from M. V. Novotny and D. Ishi, Eds., *Microcolumn Separations,* Journal of Chromatography Library, Vol. 30, Elsevier-North Holland, Inc. New York, 1985.) Reproduced with permission from Ref. (8).

detector flow cell, the optimal column configuration would have 1–2 mm ID with column length of 10–25 cm, depending on the pressure limit of the instrument. Figure 6.1 shows the various types of microcolumns including packed capillary and open tubular columns. In Table 6.3, some analytical HPLC column features are outlined. However, some of these columns are not yet commercially available. Thus their routine clinical application for drug quantitation must await further study.

1.2. Selected Literature Review

Horvath, Preiss, and Lipsky (16) were credited with the first application of small-bore columns with pellicular packings. The theoretical and practical considerations, applications, and a summary of commercially available instrumentation were succinctly summarized by Scott (15) in a recently published review article. According to Snyder's book review of that chapter (17), Sternberg's contribution (18) to the extracolumn effects should have been included. In addition, Snyder noted that reduced solvent consumption represented only a small percent of the total operating cost, and that disadvantages included problems associated with instrumentation miniaturization and with loss of concentration sensitivity. He concluded

Table 6.3. Analytical HPLC Column Features[a]

Column	Tubing	Particle Diameter (μm)	Length	Internal Diameter	Flow Rate	Commercial Availability
Conventional and narrow- bore columns	Stainless steel, glass, polyethylene	3–10	5–30 cm	2–6 mm	0.2–5 mL/min	Yes
Microbore columns	PTFE (glass- lined stainless steel)	3–30	10–50 cm	0.5–1.5 mm	10–200 μL/min	Yes
Packed capillary columns	Glass	10–100	2–50 m	40–200 μm	0.5–10 μL/min	No
	Fused silica	3–10	0.2–4 m	100–400 μm	0.5–10 μL/min	No
Open tubular columns	Glass or fused silica	—	3–20 m	10–60 μm	0.5–10 μL/min	No

[a] From Ref. (8).

that MBLC would not replace conventional HPLC for most applications, except in selected cases. Two recently published volumes of the Journal of Chromatography Library, Volumes 28 (7) and 30 (8), are both masterly treatises on microcolumn HPLC. The first, edited by P. Kucera, might be regarded as the signal of a "new era in chromatography"—a collection of theoretical and practical articles by experts from industries and universities. Of interest to clinical chemists would be the following discussions: Guiochon and Colin's discussion on the properties of different columns and some instrumentation specifications; Kucera's discussion on the design of a microbore column; Hartwick and Dezaro's application of high-speed analyses of theophylline and diazepam (theophylline was analyzed in 8 s (7); Kucera and Hartwick's applications of microbore for pharmaceuticals, biochemicals, and nucleic acids; and Henion's review of the interfacing of microbore with MS, showing the possible MBLC/MS identifications of dexamethasone, cortisone, sulfadimethoxine, sulfamethazine, some neuroleptics, betamethasone, and other biomedical and biochemical applications. Novotny and Ishii (8) published a collection of papers presented at the 1982 Honolulu meeting on microcolumns. The four parts were (1) columns, open tubular, packed capillary, and size exclusion; (2) miniaturization, highlighted by a discussion on capillary supercritical fluid chromatography; (3) spectrometry, laser-based, infrared, flame-based, and MS; and (4) electrochemical detection. In addition to the two recognized advantages, other advantages would include enhanced efficiency, possible applications of "exotic" mobile phase owing to low solvent consumption, and novel detection methods.

The following discussion is arranged according to the instrument components "downstream" from the pump, and followed by biomedical microbore analyses. For a detailed discussion on columns, the reader is referred to Scott (15) and others (7, 8). Tehrani (20) emphasized enhanced mass sensitivity and resolution through elimination of extracolumn dispersion, in combination with a small flow cell (0.5 μL) and a long optical path, and a "pulseless" syringe pump. Detection of paraben was achieved in the picogram range. The principles, practical approaches, and instrumentation for gradient elution were discussed by Schwartz, Karger, and Kucera (21), Schwartz and Berry (22), and Schwartz and Brownlee (23). Adaptation of conventional HPLC as well as dedicated systems were examined, followed by assessment on the solvent-compressibility effect, which affects the composition and flow accuracy, and the use of various high-pressure, low-volume mixers. Hayes et al. (24) demonstrated gradient MBLC/MS interfacing through the use of spray deposition of the eluent onto a moving belt, and concluded that the extracolumn dispersion was small, and the loss of resolution would be insignificant. Bowermaster and McNair (25) showed two systems for temperature programming, resulting in reduced analysis times, increased peak capacity, and selectivity control. Silver et al. (26) described microbore HPLC analysis of amino acids with a large volume injection of 20 μL (as compared to 0.5–2 μL) with a "noneluting" solvent, and showed enhanced resolution compared to analysis performed with mobile phase injection. Nielen et al. (6) showed the technical feasibility of a miniaturized precolumn for the analysis of etopside. Joshua and Schwartz (27) modified a Waters WISP injector for submicroliter injections of 0.4–40 μL with 0.1-μL increments. Cooke, Olsen, and Archer (28) compared sample detectability by UV detectors for microbore and conventional HPLC. Detection limits for 4.6- and 1.0-mm columns, for both 8-chlorotheophylline and caffeine, were 0.42 and 0.05 ng, respectively—an eightfold difference. However, the authors were critical about the general application of MBLC with UV detection. The feasibility of MBLC/MS interfacing was demonstrated (21, 29–31) by spraying eluent onto a moving belt surface (21), a fused-silica capillary interface (30), and a postcolumn liquid–liquid extraction system (31). Taylor et al. (32, 33) and Conroy et al. (34) detailed the experimental considerations of microbore HPLC/FTIR. Kucera and Umagat (35) discussed the design of a postcolumn fluorescence derivatization system, using a zigzag open tubular capillary with a special 30-nL volume mixing chamber.

Biomedical and biological microbore analyses were demonstrated for theophylline and diazepam (7, 19), pharmaceutical and various biochemicals (7), etoposide (6), corticosteroids (29), steroids (36), cardiac glycosides (37), nucleobases, nucleosides, and nucleotides (38), and phenylthiohydantoin amino acids (15, 26, 39). My preliminary study of microbore clinical assay of theophylline and caffeine (5) is outlined below in Section 1.4.

1.3. Instrumentation Survey

Instrumentation for microbore HPLC may be broadly classified into two major categories: dedicated and modified. It may be fair to state that the majority of the instrument companies offer some form of microbore hardware. The following is a brief summary based on Scott's article (15), the two volumes on microcolumns (7, 8), Schwartz and Berry's article (22), and my own experience. Owing to the "newness" of the technique, dedicated microbore systems are presently available from only a few companies: LDC/Milton (Riviera Beach, Florida), Brownlee Labs (Santa Clara, California), ISCO (Lincoln, Nebraska), and EM Science (Gibbstown, New Jersey). Although the first three suppliers utilized "pulseless" syringe pump with reservoir capacities of 5 to 50 ml, the last-named supplier's pump is a piston pump. All are capable of a flow rate of 10–1000 μL/min. The following suppliers offer modified conventional pumps (new pump head and electronics may be needed) for low flow rate: Waters, Hewlett-Packard, Kratos, Schimadzu, Jasco, LKB, and Varian. Injectors, available from Valco and Rheodyne, are capable of injecting accurately volumes of 0.5 μL or more. The WISP injector may be modified for autoinjections (27). Columns, packed with 3- or 5-μm particles with various functionalities, 1- or 2-mm ID and 100–500 mm long, are available from most of the above-listed suppliers. Other column suppliers include Anspec (Ann Arbor, Michigan), Alltech Assoc. (Avondale, Pennsylvania), Chromopack, Inc. (Bridgewater, New Jersey), Keystone Sci. (State College, Pennsylvania), and Whatman Chemical (Clifton, New Jersey). As with conventional HPLC drug analysis, the most popular detection mode is UV. Small volume-flow cells (0.5–3.0 μL) and the necessary electronics are offered by the majority of the above companies. At present, there seems to be an adequate number of suppliers to meet most of the needs related to clinical microbore HPLC drug assays.

1.4. Clinical Analysis of Theophylline, Caffeine, Procainamide, N-Acetylprocainamide, and Chloramphenicol

As a feasibility study, MBLC assays of theophylline, caffeine, procainamide, and N-acetylprocainamide were recently published by Wong et

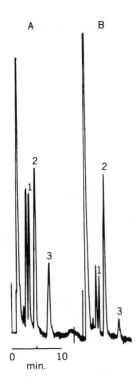

Figure 6.2. MBLC chromatograms of serum extracts of (A) 10 mg/L standard, and (B) a patient. (TH = 4 mg/L and CA = mg/L.) Peak identification: (1) TH, (2) internal standard, and (3) CA. Ref. (5).

al. (5). The emphasis was on the technical aspects, including sample preparation and practical clinical experience of operating a MBLC. For the analysis of theophylline and caffeine, protein precipitation was carried out by using 10% trichloroacetic acid. Microbore LC analysis was performed by using a 3-μm, C_{18} column (1 mm ID and 10 cm long) packed with Spherisorb™ particles. The mobile phase was 0.05 M phosphate, pH 5.0/acetonitrile (93:7), delivered at 80 μL/min. The detection wavelength was 280 mm, at 0.002 absorbance units full scale (AUFS). Figure 6.2 shows the MBLC analysis of theophylline (TH) and caffeine (CA). For the MBLC analysis of procainamide (PA) and N-acetylprocainamide (NAPA), extractions were performed by using methylene chloride. The extract was reconstituted with the mobile phase, 0.025 M phosphate, pH 3/acetonitrile (9:1) and analyzed with a similar column as indicated above. Flow rate was 100 μl/min, and the detection wavelength was 254 mm, 0.005 AUFS. Figure 6.3 shows the MBLC chromatograms. From these studies, the clinical MBLC assays of theophylline, caffeine, procainamide, and N-acetylprocainamide required careful sample preparation, and only minimal personnel retraining. This experience was transferred to

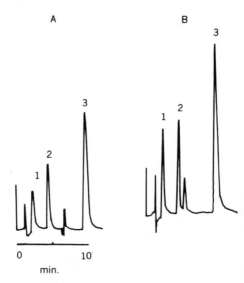

Figure 6.3. MBLC chromatograms of serum extracts of (A) 5 mg/L standard, and (B) a patient (PA = 7 mg/L and NAPA = 5 mg/L). Peak identification: (1) PA, (2) NAPA, and (3) internal standard. Ref. (5).

develop a MBLC assay of an antimicrobial, chloramphenicol, because this drug is preferentially assayed by LC. In order for the reader to appreciate the analytic strategy, the clinical pharmacology of chloramphenicol is discussed briefly, followed by some relevant analytic considerations.

Chloramphenicol, initally extracted from the bacterium *Streptomyces venezuela* in 1947, is a broad-spectrum antimicrobial, active against gram-positive and gram-negative organisms and rickettsiae (40–45). It inhibits protein synthesis by binding to the 50S ribosomal subunit in bacterial walls. In pediatric cases, chloramphenicol may be used for the treatment of meningitis and epiglotitis caused by *Haemophilus influenzae*. Adverse reactions of chloramphenicol include aplastic anemia, possible bone marrow suppression, and "gray" syndrome. The therapeutic range is proposed to be 10–25 mg/L.

Chloramphenicol may be administered intravenously (IV) or orally. For neonatal and pediatric patients, the preferred route is IV for the prodrug chloramphenicol succinate esters. Figure 6.4 shows the chemical structure of the succinate esters and the conversion in the liver into active drug chloramphenicol. Note that the prodrug esters exist in a ratio of 80:20 as the 3- and 1-esters under physiological conditions (46). For oral

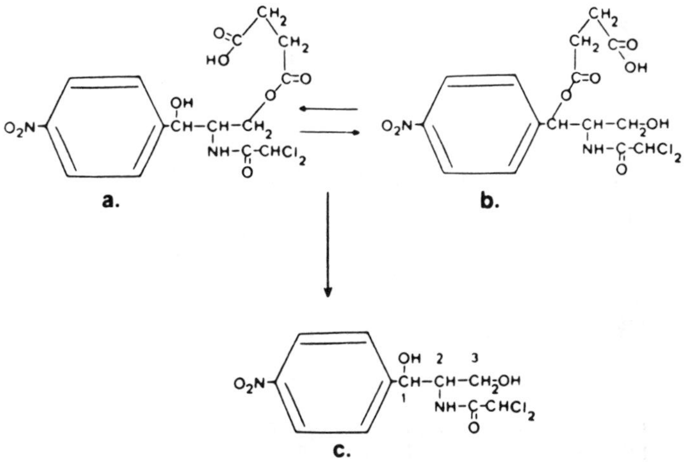

Figure 6.4. Chemical structures of (*a*) chloramphenical-3-succinate, (*b*) chloramphenicol-1-succinate, and (*c*) chloramphenicol. Under physiological conditions, (*a*) and (*b*) equilibrate to yield an 80:20 mixture of isomers, respectively. Conversion of (*a*) and (*b*) to give the active drug chloramphenicol (*c*) occurs primarily in the liver. Ref. (41).

administration, chloramphenicol palmitate is usually given as a liquid suspension. It is hydrolyzed by pancreatic lipases in the duodenum. Chloramphenicol is then absorbed across the gut, peaking in about 2 hr postingestion. Thus, for the LC analysis of chloramphenicol in pediatric and neonatal patients, it may be necessary to resolve chloramphenicol from the prodrug esters and to have adequate sensitivity. These requirements have been reliably achieved by published LC procedures and more recently by an enzyme multiplied immunoassay technique (EMIT). Using MBLC, because of enhanced sensitivity, as little as 5 µL of serum would be sufficient. This would allow additional drug or chemistry assays from a single microtainer or Natelson tube, which is capable of collecting up to 100 µL of whole blood.

In designing the MBLC assay, the preliminary study showed that a conventional LC assay procedure may be readily adapted, if chloramphenicol is the only drug peak, unaccompanied by the prodrug esters peaks, as in serum from patients medicated with oral doses of chloramphenicol palmitate. However, for serum from patients given chloramphenicol succinate esters, the LC assay overestimated the amount of chloramphenicol, indicating that the esters co-eluted with chloramphenicol. Extensive solvent scouting experiments led to the use of a ternary mobile phase, using tetrahydrofuran (THF), in combination with elevated temperature. This approach did resolve the prodrug peaks, but required

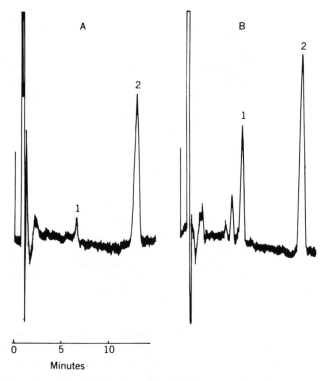

Figure 6.5. Microbore liquid chromatograms of 0.5-μL aliquots of supernatant from procedure B. Chromatogram A shows 6 mg/L of chloramphenicol following oral administration of chloramphenicol palmitate, and chromatogram B shows 21 mg/L of chloramphenicol following IV administration of chloramphenicol succinate. Peak identification: 1 = chloramphenicol, 2 = internal standard). Ref. (9).

lengthy equilibration, probably due to the elevated temperature. Because the attenuation was set at a very sensitive scale of about 0.005–0.002 AUFS, any nonequilibration resulted in a drifting base line, rendering peak height measurement difficult. This difficulty led to the investigation of another approach—increased interaction of analytes with a high carbon load column. Based on our previous experience with antidepressant monitoring with a column with high carbon load—20% instead of 10%—we thought this column might offer the needed, enhanced selectivity for the separation of structurally similar analytes such as the 3- and 1-succinate esters. Indeed, the separation was achieved at ambient temperatures as shown by Figure 6.5. The finalized protocol, utilizing only 5 μL of serum, included a protein precipitation step by mixing with 20 μL of methanolic solution containing an internal standard, 5-ethyl 5-*p*-tolylbarbituric acid. Other important technical considerations include extended high-speed (10,000 × g) centrifugation to yield a well-defined clear, upper layer, and careful sampling of the supernatant to avoid injection of particles. The assay was found to be precise, and it compared favorably to the conventional LC assay. It may be used to quantitate samples from patients who had been given either chloramphenicol succinate (IV) or palmitate (oral), and is especially useful for limited-sample-size analysis such as in the case of pediatric and neonatal samples.

2. DIRECT-SAMPLE ANALYSIS

Because chromatography, LC in particular, may allow simultaneous determination of drugs and/or metabolites, and may be more cost-effective when compared to immunoassay, it has been used for drug analysis in the clinical laboratory. In particular, direct-sample analysis capitalizes on the above and the following additional advantages: the lack of manual sample preparation, which renders automation feasible; the minimal exposure of infectious sample to the analyst, a most noteworthy consideration in view of the AIDS virus; increased efficiency and precision; and analysis of light-sensitive drugs and metabolites. The current approaches all involve components "downstream" from the pump as follows: (*a*) "precolumn," including extraction column (Advanced Automated Sample Processor™, AASP) (50), solvent extraction (51), and column switching (6); (*b*) "on-column," including microinjection (10, 11, 14), micellar chromatography (12), and the bimodal, internal surface reverse phase (13); and (c) "postcolumn" photolytic derivatization. All these approaches involve manipulation of the sample both prior to and after the analysis. Because the majority of these techniques utilize a UV detector, the most common detection mode in clinical and research laboratories, the detectability and sensitivity requirements dictate that these drugs should be UV absorbing and in the concentration ranging from milligrams per liter to micrograms per liter, depending on the sample size. The following outlines representative examples of each of these techniques. Also, a discussion of bimodal, internal surface reverse phase liquid chromatography is presented here, based on my experience.

2.1. Advanced Automated Sample Processor (AASP)

Ni et al. (50) demonstrated the use of a commercial, automated on-line extraction system AASP for the analysis of tricyclic antidepressants.

After conditioning of the cartridge containing C_8 packing, aliquots of plasma, internal standard solution, and phosphate buffer were introduced into the reservoir. After mixing, aliquots were introduced onto the cartridge and phosphate buffer (pH 11.5) was used for cleanup. Through programmed backflushing with mobile phase, automated analysis was achieved with an alkylnitrile bonded column. Analysis time of each sample was about 10 min.

2.2. Solvent Extraction

Pioneered by Snyder and co-workers, this dedicated "FAST-LC" for drugs (51) applied Technicon's clinical chemistry technology for extraction. Typically, serum was loaded into containers/curvettes in a tray as shown in Figure 6.6. Successive aliquots were then pipetted into the mixing and extracting coils. Further extraction may be carried out, followed by injection into the LC. Such a system was developed for antiepileptics, antidepressants, and other drugs. Unfortunately, "FAST-LC" has never been used for routine clinical drug assays.

2.3. Column Switching

Frei and co-workers utilized this technique for the analysis of an anticancer drug, etoposide (6). The principle is based on on-line trace enrichment, followed by narrow-bore LC analysis. Figure 6.7 shows the instrumentation. After loading aliquots (100 μL) of serum, trace enrichment was achieved on the micro-precolumn by using water as the mobile phase. Selected injection was then carried out by eluting the micro-precolumn with the mobile phase, followed by analysis with the analytical column. As shown in Figure 6.7, two LC pumps were used in series. The authors proposed that the precolumn may be used for loading serum such as field samples in doctors' offices, followed by LC analysis in the laboratory.

Nazaki (52) demonstrated the measurement of acetaminophen with instrumentation similar to that of Frei and co-workers: two columns, two pumps, and a switching valve. The first column was a BSA–ODS column, eluted with 5% methanol in 7 μM phosphate. After 10–20 μL of serum mixed with the internal standard, o-acetamidophenol, was injected, a selected portion was injected into an analytical column, Nucleosil, ODS. Retention time of acetaminophen was 6.5 min. Nazaki found that the procedure was precise and linear (2–120 μg/mL) and provided good recovery.

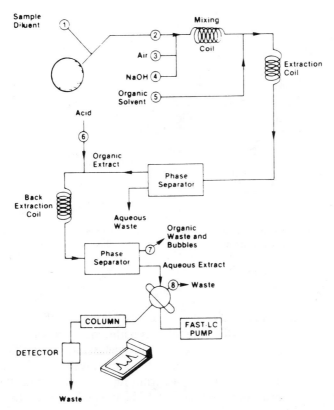

Figure 6.6. Flow diagram of sample pretreatment module used with FAST-LC system. Ref. (51).

2.4. Microinjection

Shihabi et al. (10) recently demonstrated the analysis of pentobarbital by injecting 2 μL of serum into a polymeric reverse phase (PRP-1) column, as shown in Figure 6.8. Serum precipitation was minimized by injecting small 2-μL aliquots at a highly basic pH of 11.8. The polymeric column was stable even after 300 injections. The authors suggested that the procedure is amenable to automation and STAT assays. In addition to the polymeric column, Shihabi et al. (11) explored the use of wide-pore particles, 300 Å, and low-hydrophobicity silica columns. Again, small, 5-μL samples were injected for the analysis of carbamazepine. The column was stable after 300 injections. The method compared favorably with fluorescence polarization immunoassay (FPIA).

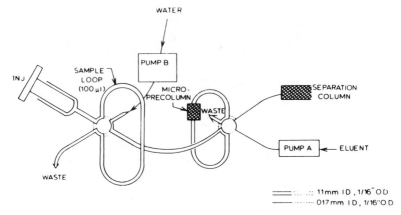

Figure 6.7. Direct plasma injection and on-line trace enrichment of etoposide. Precolumn 5 × 1 mm ID. Analytical column, 30 cm × 1 mm ID packed with 5 μm LiChrosorb RP-18. Eluent, 40% acetonitrile at 100 μL/min. Pump B, 200 μL/min. Detection by fluorescence at 350 (ex.)/328 (em.) nm. Ref. (6).

Acetaminophen was analyzed in serum by diluting with internal standard solution without further sample preparation. Typically, 10 μL of serum was mixed with 0.5 mL of the internal standard, 8-chlorotheophylline solution. Then 20 μL was injected for analysis on a C_{18} column. With this procedure, the amount of serum injected was about 0.02 μL. Even though the procedure had proved to be cost-effective, reliable, and precise, it was recently replaced by a much faster procedure using FPIA in my laboratory. This decision was based upon the relative small number of clinical samples, and after-hour requests (times other than 8:30 AM to 12:00 AM).

2.5. Micellar Chromatography

DeLuccia et al. (12) have used surfactant mobile phases such as anionic SDS and nonionic Brij-35 to achieve drug sample analysis. The principle is based on the following proposed reactions among micelles, monomers, serum proteins, and drugs. The surfactant monomers, at about the critical micellar concentration, would bind to insoluble serum components and would solubilize them as a result of the charged/polar coating. Free monomers and/or micelles may displace the protein-bound drugs. During the analysis, the serum protein micelle complexes elute right after the solvent front, whereas the elution of the drug/micelle/monomers would depend upon the SDS micelle concentration. Figures 6.9 and 6.10 show the anal-

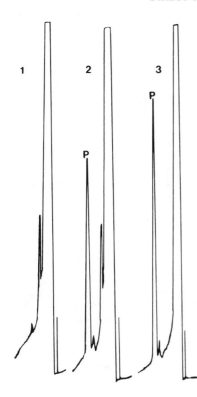

Figure 6.8. Representative chromatogram of (1) patient serum free from pentobarbital, (2) patient treated with pentobarbital, and (3) aqueous standard 50 mg/L (P = pentobarbital peak; time for P = 6 min). Serum (2 μL) was injected directly onto the column and eluted as described with detection at 254 nm. Ref. (10).

ysis of theophylline and carbamazepine. The sensitivity of this procedure may be enhanced by selective fluorescence, such as in the analysis of quinidine. This would minimize background noise, and eliminate any solvent-front protein–micelles complex peaks.

2.6. Electrochemical Detection with Photolytic Derivatization

Selavka and Krull (14) demonstrated the use of LCEC for drug analysis with morphine, narcotics, benzodiazepine, cannabinoids, fentanyl, and tricyclic antidepressants. With LC–photolysis–EC, components with no inherent electroactivity would be rendered detectable. These include nitrocompounds, beta-lactam, morphine, nalorphine, pethidine, cannabinoids and metabolites, cocaine, benzoylecgonine and ecgonine, barbiturates, methylphenidate, lysergic acid diethylamide, benzodiazepines, and others.

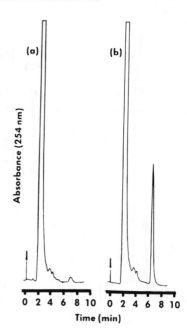

Figure 6.9. Chromatograms of (*a*) serum blank and (*b*) serum with 10 μg/mL theophylline added. Ref. (12).

2.7. Bimodal, Internal Surface Reverse Phase

Pinkerton (13) introduced a new concept, termed internal surface reverse phase (ISRP), a unique mode of multimodal separation. ISRP may be characterized as a bimodal separation, whereby the reverse phase mode is carried out by the hydrophobic functional group, a peptide bound to the internal surface of the porous silica, and the size exclusion mode is carried out by the small-pore silica packing (pore size of 52 Å). Serum protein molecules with molecular weight of >5000 are excluded and eluted close to the solvent front peaks. Adsorption of protein is minimized by the external, hydrophilic glyceryl-L-phenylalanine-L-phenylalanine. The chemistry and characteristics of this novel packing have been thoroughly characterized by Pinkerton.

Because the bimodal ISRP may achieve micro-sample analysis of drug in serum without sample preparation, it may be easily automated with sample injector and data processing capability, both controlled by computer. And it has potential clinical applications that include (*a*) neonatal and pediatric drug monitoring; (*b*) light-sensitive drug analysis; and (*c*) toxicology screening and confirmation testings. My laboratory has begun a systematic feasibility study for the neonatal and pediatric drug moni-

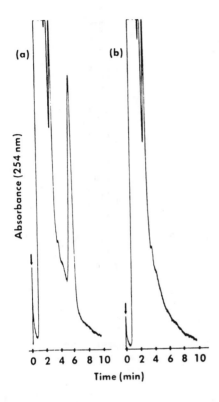

Figure 6.10. Chromatograms of (*a*) serum with 10 μg/mL carbamazepine added and (*b*) serum blank. Ref. (12).

toring by monitoring carbamazepine and comparing the measurements with those of an established immunoassay (53).

The ISRP column used was 25-cm long, coupled with a guard column. The mobile phase was 0.1 M phosphate, pH 6.8, isopropanol, and THF (86:10:4), delivered at 1.0 mL/min. Detection wavelength was 254 nm at 0.005 AUFS. Prior to injection of the serum, aliquots (10–100 μL) were centrifuged at 9500 g for about 10 min. Then 1-μL aliquots were injected into the column. Figure 6.11 shows the results of such analysis for carbamazepine in serum. The calibration was linear up to 20 mg/L. The ISRP assay was compared to a clinical carbamazepine (Abbott) FPIA for 20 patient specimens. Both linear regression and paired-*t* test showed that these two techniques were comparable.

3. CONCLUSIONS

As chromatography approaches its half-century anniversary, it would be appropriate to speculate that within the next 10 years, gas–liquid, liquid,

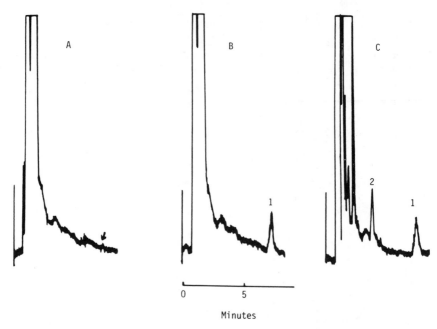

Figure 6.11. ISRP chromatograms of carbamazepine in serum: A, drug-free serum; B, 8 mg/ L standard; and C, patient serum with estimated concentration of 8 mg/L. Ref. (53).

thin-layer, and possibly supercritical fluid chromatography will play various roles in the field of TDM and toxicology. The extent of their roles will depend on the technological advances, clinical drug monitoring/identification requirements, the success of other competitive techniques such as immunoassay, and the changing drug utilization pattern of our society. The two techniques described here, microcolumn and direct-sample-analysis LC, would fulfill some of these requirements. Although they may be characterized as rapidly maturing techniques, the extent of their routine clinical application awaits further study.

REFERENCES

1. S. H. Wong, "Introduction: Status of Therapeutic Drug Monitoring," in S. H. Y. Wong, Ed., *Therapeutic Drug Monitoring and Toxicology by Liquid Chromatography*, Marcel Dekker, New York, 1985, pp. 1–10.
2. Alcohol, Drug Abuse and Mental Health Administration, Scientific and Tech-

nical Guidelines for Drug Testing Program, Department of Health Services, Washington, DC, Feb. 1987.

3. S. H. Wong, "Liquid Chromatographic and Other Methodologies for Drug Monitoring," in S. H. Y. Wong, Ed., *Therapeutic Drug Monitoring and Toxicology by Liquid Chromatography*, Marcel Dekker, New York, 1985, pp. 39–78.

4. S. H. Y. Wong, *Lab. Management, 23,* 59 (1985).

5. S. H. Y. Wong, N. Marzouk, O. Aziz, and S. Sheeran, *J. Liq. Chromatogr., 10,* 491 (1987).

6. M. W. F. Neilen, E. Sol, R. W. Frei, and U. A. Th. Brinkman, *J. Liq. Chromatogr., 8,* 1053 (1985).

7. P. Kucera, Ed., *Microcolumn High-performance Liquid Chromatography,* Journal of Chromatography Library, Vol. 28, Elsevier, Amsterdam, 1984.

8. M. V. Novotny and D. Ishii, Eds., *Microcolumn Separations,* Journal of Chromatography Library, Vol. 30, Elsevier, Amsterdam, 1985.

9. S. H. Y. Wong, B. Cudny, O. Aziz, N. Marzouk, and S. R. Sheeran, *Clin. Chem., 33,* 1021 (1987).

10. Z. K. Shihabi, R. D. Dyer, and J. Scaro, *J. Liq. Chromatogr., 10,* 663 (1987).

11. Z. K. Shihabi and R. D. Dyer, *Clin. Chem., 33,* 1018 (1987).

12. F. J. DeLuccia, M. Arunyanart, P. Yarmchuk, R. Weinberger, and L. J. Cline Love, *Liq. Chromatogr., 3,* 794 (1985).

13. T. C. Pinkerton, T. D. Miller, J. E. Cook, et al., *Biochromatography, 1,* 96 (1986).

14. C. M. Selavka and I. S. Krull, *J. Liq. Chromatogr., 10,* 345 (1987).

15. R. P. W. Scott, "Small-bore Columns in Liquid Chromatography," in J. Giddings et al., Eds., *Adv. Chromatogr., 22,* 247 (1983).

16. C. Horvath, B. A. Preiss, and S. R. Lipsky, *Anal. Chem., 34,* 1422 (1967).

17. L. R. Snyder, "Book Review—Advances in Chromatography, Vol. 22," *Liq. Chromatogr., 2,* 547, 1985.

18. J. C. Sternberg, *Adv. Chromatogr., 2,* 205 (1966).

19. D. Dezaro, D. Horn, and R. A. Hartwick, *Clin. Chem., 29,* 1158 (1983).

20. A. Y. Tehrani, *Liq. Chromatogr., 3,* 42 (1985).

21. H. E. Schwartz, B. L. Karger, and P. Kucera, *Anal. Chem., 55,* 1752 (1983).

22. H. E. Schwartz, and V. V. Berry, *J. Liq. Chromatogr., 3,* 110 (1985).

23. H. Schwartz and R. G. Brownlee, *Am. Lab., 6,* 43 (1984).

24. M. J. Hayes, H. E. Schwartz, P. Vouros, B. L. Karger, A. D. Thruston, and J. M. McGuire, *Anal. Chem., 56,* 1229 (1984).

25. J. Bowermaster and H. M. McNair, *J. Chromatogr. Sci., 22,* 165 (1984).

26. M. R. Silver, T. D. Trosper, M. R. Gould, et al., *J. Liq. Chromatogr., 7.,* 559 (1984).

27. H. Joshua, and R. E. Schwartz, *Liq. Chromatogr., 3,* 50 (1985).

28. N. H. C. Cooke, K. Olsen, and B. G. Archer, *Liq. Chromatogr.*, *2*, 514 (1984).

29. D. S, Skrabalak, T. R. Covey, and J. D. Henion, *J. Chromatogr.*, *315*, 359 (1984).

30. P. Hirter, H. J. Walther, and P. Datwyler, *J. Chromatogr.*, *323*, 80 (1985).

31. A. Apffel, U. A. Th. Brinkman, and R. W. Frei, *J. Chromatogr.*, *312*, 153 (1984).

32. R. S. Brown and L. T. Taylor, *Anal. Chem.*, *55*, 1492 (1983).

33. P. G. Amateis, and L. T. Taylor, *Liq. Chromatogr.*, *2*, 854 (1984).

34. C. M. Conroy, P. R. Griffiths, and K. Jinno, *Anal. Chem.*, *57*, 822 (1985).

35. P. Kucera and H. Umagat, *J. Chromatogr.*, *255*, 563 (1983).

36. M. Novotny, M. Alasandro, and M. Konishi, *Anal. Chem.*, *55*, 2375 (1983).

37. Y. Fujii, R. Fujii, and M. Yamazaki, *J. Chromatogr.*, *258*, 147 (1983).

38. S. Robushika, Z. Y. Oii, and M. Hatano, *J. Chromatogr.*, *330*, 335 (1985).

39. R. L. Cunico, R. Simpson, L. Ocrreia, and C. T. Wehr, *J. Chromatogr.*, *336*, 105 (1984).

40. D. T. Mines, "Antibiotics," in S. H. Y. Wong, Ed., *Therapeutic Drug Monitoring and Toxicology by Liquid Chromatography*, Marcel Dekker, New York, 1985, pp. 269–307.

41. K. E. Opheim, "Chloramphenicol Succinate: A Pro-drug Form of Chloramphenicol," in *Therapeutic Drug Monitoring Continuing Education and Quality Control Program, American Association for Clinical Chemistry*, Vol. 5, No. 9, pp. 1–4, 1984.

42. H. C. Meissner and A. L. Smith, *Pediatrics*, *64*, 348 (1979).

43. P. S. Lietman, *Clin. Perinatol.*, *6*, 151 (1979).

44. B. M. Goldsmith and R. L. Boeckx, "Chloramphenicol Toxicity Associated with Hepatic Dysfunction," *Therapeutic Drug Monitoring Continuing Education and Quality Control Program, American Association for Clinical Chemistry*, Washington DC, April, 1983.

45. A. J. Glazko, H. E. Carnes, A. Kazenko, L. M. Wolf, and T. F. Reutner, *Antibiot. Ann.*, *1958*, 792 (1957–1958).

46. D. A. Brent, P. Chandrasurin, A. Ragouzeos, B. S. Hurlburt, and J. T. Burke, *J. Pharm. Sci.*, *69*, 906 (1980).

47. J. T. Burke, W. A. Wargin, and M. R. Blum, *J. Pharm. Sci.*, *69*, 909 (1980).

48. M. K. Aravind, J. N. Miceli, R. E. Kaufman, L. E. Strebel, and A. K. Done, *J. Chromatogr.*, *221*, 176 (1980).

49. S. H. Petersdorf, V. A. Raisys, and K. E. Opheim, *Clin. Chem.*, *25*, 1300 (1979).

50. P. Ni, F. Guyon, M. Caude, and R. Rosset, *J. Liq. Chromatogr.*, *11*, 1087 (1988).

51. S. J. Bannister, S. van der Wal, J. W. Dolan, and L. R. Snyder, *Clin. Chem.,* *27,* 848 (1981).

52. O. Nozaki, T. Ohata, and Y. Ohba, *Clin. Chem., 33,* 1026 (1987).

53. S. H-Y. Wong, B. Cudny, and S. Sheeran, *Ann. Clin. Lab. Sci., 17,* 271 (1987).

CHAPTER

7

HIGH PERFORMANCE LIQUID CHROMATOGRAPHY/ MASS SPECTROMETRY

ROBERT D. VOYKSNER

Research Triangle Institute
Research Triangle Park, North Carolina

1. INTRODUCTION

Mass spectrometry is recognized as an important tool for acquiring both qualitative and quantitative data for many biologically interesting molecules. An analysis by mass spectrometry (MS) can provide molecular weight and isotope ratios, and aid in structure elucidation through fragmentation patterns. Mass spectrometry also serves as a selective, sensitive detection method for target compounds in complex biological matrices. For those not familiar with the capabilities or hardware of mass spectrometry, a general text by Watson (1) would prove informative.

Initial applications of MS to clinical problems were made on relatively pure samples that could be vaporized in the ion source and ionized by electron impact (EI). The coupling of gas chromatography with MS (GC/

MS) enabled the study of minor components in complex mixtures and removed purity constraints from an analysis. Through chemical derivatization a large variety of compounds lacking volatility have been analyzed without decomposition. However, derivatization has often been difficult, time consuming, and at times, unsuccessful, especially in trace analysis application. The development of chemical ionization (CI) enabled the analysis of compounds that do not produce molecular ions (M) by EI conditions in GC/MS. CI is a soft ionization technique in which the analyte is ionized through gas phase reactions with a reagent gas, usually resulting in $[M + H]^+$ ion formation. Although CI can prove useful in solving analytical problems, the technique requires the sample to be in the gas phase, limiting its usefulness for the analysis of nonvolatile compounds. Recent developments in new desorption ionization techniques have overcome the need to volatilize a sample for MS analysis and have proved gentle toward most compounds, thus enabling the observation of molecular ions. These ionization techniques include fast atom bombardment (FAB) (2), field desorption (FD) (3), and laser desorption (4). Nonvolatile and thermally labile compounds can be separated by high performance liquid chromatography (HPLC), and these new desorption ionization techniques permit their detection by MS.

The emergence of HPLC in recent years has been prompted not only by the technical advances of bonded phases and high resolution, but also by the fact that only 20–30% of the known organic compounds can be analyzed by GC techniques without derivatization. HPLC techniques can be applied to a much broader range of compounds including thermally sensitive, polar, and high-molecular-weight compounds for which GC is not applicable. The range of detectors available for HPLC is not as diverse or usually as sensitive as those available for GC. Refractive index (RI) detectors are the most "universal" or nonspecific used for HPLC, much as flame ionization detection (FID) is for GC. The ultraviolet (UV) absorbance detector for HPLC is the most common and is a specific detector. This detector is limited by the solvent's UV absorption and the necessity that the compounds have chromophores. Other detectors include electrochemical, fluorescence, and photoionization. Some of these detectors offer increased selectivity and sensitivity relative to UV detection, but only for compounds with specific functional groups or structural characteristics. A detector that could be selective and sensitive to a much broader range of compounds would be very desirable.

Clearly, the combination of HPLC with MS detection could prove to be as powerful an analytical tool as the widely accepted technique of GC/MS. However, the coupling of HPLC and MS suffers from basic incompatibilities between the two techniques. First, the MS operates at a low gas pressure (about 10^{-6} torr) whereas the HPLC operates at atmospheric pressure at flows of 0.5–2.0 mL/min of liquid. These HPLC flows correspond to about 800–to 3800 mL/min of gas (quantity of gas produced depends on the solvent choice), which is nearly 100 times the amount that can be handled by the MS. Second, suitable ionization techniques for MS that do not require thermal vaporization of the sample must be used. Although techniques such as EI or CI are suitable for numerous compounds analyzed by HPLC, these conventional ionization techniques are limited to gas-phase ionization processes that are not suitable for the nonvolatile or thermally labile compounds typically analyzed by HPLC. Clearly, the combination of HPLC and MS will require an interface that compensates for the operating incompatibilities while not compromising the chromatographic performance or MS operation.

This chapter reviews the current state of combined HPLC/MS. Included in this review are numerous examples of HPLC/MS applications to solving problems relevant to clinical toxicology, specifically applications of HPLC/MS in the analysis of steroids, alkaloids, mycotoxins, drugs, and pesticides. The use of HPLC with tandem MS (HPLC/MS/MS) and quantitative application of HPLC/MS also are covered.

2. HPLC/MS INTERFACES

Many different approaches have been attempted in combining HPLC and MS. The merits and disadvantages of most of these approaches have been extensively reviewed (5–10). Most introductory work in HPLC/MS focused on the first problem, incompatibility in flow acceptance between the two techniques. Recently, more research has been devoted to the second problem, developing ionization techniques that are compatible with most organics and do not require sample volatility. This section discusses current interfaces that show promise and are commercially available for potential widespread use.

The approaches of combined HPLC/MS can be divided into four categories: off-line techniques, on-line approaches of direct introduction, direct introduction with enrichment, and mechanical transport.

2.1. Off-Line HPLC/MS

Off-line HPLC/MS, once popular when on-line HPLC/MS coupling was considered unapproachable, is rapidly diminishing in use. Off-line methods involve collection of HPLC fractions, evaporation of the HPLC solvent, and transfer of the concentrated residue to a solid probe for MS analysis. This method totally isolates the HPLC from the MS, thus allowing for independent operation of each instrument. The MS can be operated in nearly any ionization mode (EI, CI, FAB, FD, etc.) to obtain the spectrum of the analyte (4, 11). Although this flexibility is appealing, it is outweighed by the tedious, time-consuming work of fraction collections and solvent evaporation. Collecting of fractions every minute of a 30-min analysis of a biological sample in which new metabolites are sought could easily require an entire day to analyze. Also, each fraction could contain three to seven components that could easily interfere with the MS analysis. Therefore, the loss in HPLC resolution in fraction collection makes identifications of unknowns or of very narrow peaks nearly impossible. Secondly, off-line techniques require the use of a conventional HPLC detector, which has intrinsic limitations, making it difficult to detect new compounds or metabolites that lose their chromophore. Except for a few unique cases, most HPLC/MS work could be performed better on-line, as is GC/MS.

2.2. Direct Introduction

Direct introduction is the simplest approach to combined HPLC/MS. Basically, direct introduction interface would allow the maximum amount of HPLC effluent into the MS source without hindering the MS performance. An early interface developed by Tal'Roze et al. (12) allowed nanoliter per minute quantities of effluent to enter the MS source, enabling acquisition of EI type spectra. However, this reduced flow posed severe limitations on the HPLC and MS sensitivity. Baldwin and McLafferty (13) and McLafferty et al. (14) demonstrated that direct coupling to a CI instrument would allow three orders in magnitude more HPLC flow into the MS, resulting in microliter-per-minute flows entering the MS source. Currently, most direct introduction interfaces operate in the CI mode. Ionization of the sample occurs from solvent-derived ions generated from a filament, which undergo gas-phase reactions with the sample to generate sample ions. These ions, as well as the HPLC solvent reagent ions, are extracted from the source and mass analyzed. The nature of the sample ions detected depends on the reactivity and concentration of the reagent ions and the thermodynamics of the source. Primarily, $[M + H]^+$ ions are generated in positive ion operation because most polar solvents are strong gas-phase acids that protonate the sample. Sometimes $[M + \text{solvent}]^+$ or $[M + \text{buffer}]^+$ ions are observed when the sample has a high gas-phase acidity. In the negative ion mode, $[M]^-$ ions from electron capture and $[M - H]^-$ ions from proton abstraction are usually detected. Sometimes $[M + \text{solvent}]^-$ or $[M + \text{buffer}]^-$ ions are detected. The appearance of fragment ions in a spectrum is compound-dependent and difficult to predict.

The most common and simplest method of direct coupling is direct liquid introduction (DLI) pioneered by McLafferty et al. (14) and demonstrated by Sugnaux et al. (15), Henion (16), Dedieu et al. (17), and Melera (18). The DLI interface (Fig. 7.1) was designed to be compatible with most commercial direct probe inlet systems and CI sources. A 5-μm orifice at the probe tip creates a fine liquid stream. A restrictor controls the column of this stream by the pressure drop across the orifice to enable adjustment of flow into the instrument. Typically, 5–15 μL/min flows are achievable. If cryo-cooling or larger pumps are employed, flows of 20–50 μL/min can be achieved. The DLI interface could operate with most HPLC solvents and volatile buffers including pentane, hexane, octane, chloroform, acetonitrile, methanol, tetrahydrofuran, water, and buffers such as ammonium acetate, acetic acid, ammonium hydroxide, and trifluoroacetic acid. The DLI interface is often maintained at room temperature with a water chiller to prevent evaporation of the solvent in the high-pressure side of the orifice by heat transfer from the source block. Solvent evaporation may result in an unstable ion current.

DLI–HPLC/MS has been employed extensively to solve real problems, but the technique is limited. The most severe limitation is the sensitivity loss in the 1:100 split necessary to accommodate MS operation at HPLC flows of 1–2 mL/min. The use of micro-HPLC columns coupled with MS has removed this need for a split because most micro-HPLC columns operate at conditions acceptable for DLI–HPLC/MS (10–50 μL/min). However, most separations are developed on normal columns at 1–2 mL/min, and application of these separations to DLI-HPLC/MS requires the investigator to develop HPLC conditions for micro columns. The DLI interface also suffers from occasional plugging of the 5-μm orifice and is limited to gas-phase ionization techniques under CI conditions. The plugging of the orifice is not irreversible because the diaphragm can be removed and cleaned easily, resulting only in the loss of the HPLC analysis. The limitation of the interface to volatile samples is more restricting. Some work with smaller-orifice diameters (1–3 μm) and heated desolvation chambers have enabled the analysis of nonvolatile compounds, such as adenosine (18), and thermally unstable compounds (17). Yet the range in volatility of compounds amenable to DLI has not been fully explored.

Another variation of directly coupling HPLC and MS has been the substitution of supercritical fluid chromatography (SFC) (19) for HPLC. In SFC the mobile phase is a supercritical fluid rather than a liquid. At sufficient fluid density and high pressure, the analyte is solvated as in a liquid, but mobilities are nearer those of a gas phase. Thus analyte vol-

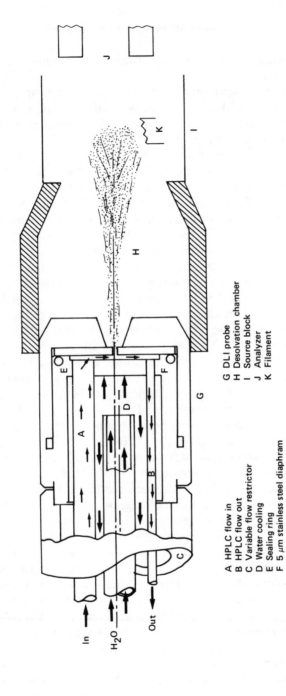

A HPLC flow in
B HPLC flow out
C Variable flow restrictor
D Water cooling
E Sealing ring
F 5 μm stainless steel diaphram

G DLI probe
H Desolvation chamber
I Source block
J Analyzer
K Filament

Figure 7.1. Schematic of a DLI interface.

178

atility is not required. The separation efficiencies and speed of SFC are similar to GC. Temperatures do not have to be high (~40°C) to achieve supercritical fluid conditions if carbon dioxide is used. Using a capillary column, SFC allows the entire effluent, which is more volatile than HPLC mobile phases, to enter the MS. Typically, a DLI interface without a split can be used for coupling SFC and MS. The SFC mobile phase acts as the CI reagent gas to ionize the sample, analogous to DLI-HPLC/MS. If an inert gas such as carbon dioxide is used, the filament will generate spectra that resemble those acquired by EI, or a reagent gas such as ammonia can be added to acquire CI spectra.

This technique is relatively new and there are only a few applications of its capabilities, presented by Smith et al. (20, 21). The available data are insufficient to determine the problems with the technique or if it will be useful for the analysis of nonvolatile compounds.

A second method of directly coupling HPLC with MS was pioneered by Carroll et al. (22), who developed an MS source that could accept very high pressures. The atmospheric pressure ionization (API) source (Fig. 7.2) allows the entire HPLC effluent to enter the MS. The HPLC effluent is vaporized by a heated stream of nitrogen, and the vapor is ionized by a electrical corona discharge. The discharge produces reagent ions from the HPLC solvent for ionization of a sample. A fraction of the ions in the source is sampled through a 25-μm orifice into the MS while most of the gas vapor is kept out of the vacuum. Despite promising preliminary work little development or use of the technique has been shown. Recent developments such as nebulization and liquid ion evaporation ionization have renewed interest in the technique. Now, API has the ability to separate the nebulization and ionization processes (Fig. 7.2) (rather than combine them as in thermospray, discussed below), which can aid the ionization of nonvolatile compounds (23, 24).

The API system has several drawbacks: (1) Ionization occurs in the gas phase; therefore the compounds must be volatilized before they can be ionized and analyzed by MS. (2) Trace impurities or the choice of solvents can reduce the reagent ion concentration, thereby reducing sensitivity. (3) Volatile buffers must be used in the analysis to prevent plugging of the sampling orifice. (4) Finally, the high pressure in the source causes the HPLC solvent (especially polar solvent) to cluster, yielding $[(\text{solvent})_n + \text{H}]^+$ ions ($n = 1$–5). These ions can interfere with the sample analysis and overwhelm the low-mass section of the spectra. However, some of these problems have been overcome recently through the use of nebulization.

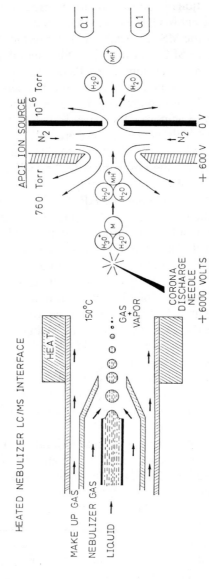

Figure 7.2. Schematic of an API source with heated nebulizer for HPLC/MS. Reprinted with permission from T. Covey, E. Lee, A. Bruins, J. Henoin *Anal. Chem. 58*, (14) 1456A. Copyright 1986, American Chemical Society.

2.3.   Direct Introduction with Enrichment

Enrichment processes overcome problems of having to discard large por-
tions of the HPLC effluent and increase the probability of ionizing and
analyzing the sample relative to the HPLC solvent. In packed-column
GC/MS, jet separators often perform a 10–20:1 enrichment of the sam-
ple:carrier gas, but for HPLC/MS, the enrichment may have to exceed
1000:1. The permeability differences between the mobile phase and the
analytes are much lower than in GC/MS, making enrichment by selective
permeability very difficult. Two approaches to enrich the sample yield at
normal HPLC flow rates have been attempted.

The first approach, developed by Jones and Yang (25), used a three-
stage silicone rubber membrane to concentrate a nonpolar sample separated using polar solvents. The work presented was novel, but has limited
practical use because of the demands on the polarity of the solvents and
sample to achieve enrichment.

The second approach is thermospray HPLC/MS developed by Vestal
and Blakely et al. (26). The thermospray interface (Fig. 7.3) evolved from
its initial use of lasers (27), to the oxy-hydrogen flame (28), then to electrical heating (29), which presently is used to vaporize and ionize the
HPLC effluent. The supersonic stream formed from the vaporizer enters
the source where ions in this stream are extracted through the ion exit,
and the neutral molecules continue to a cold trap and mechanical pump.
This extraction process allows for the total introduction of 1–2 mL/min
of HPLC effluent while an MS pressure of 10^{-5} torr is maintained.

Thermospray is an ionization technique as well as an enrichment technique. Ions may be produced by conventional CI methods, initiated by
a filament or discharge, or through ion evaporation. Ion evaporation is
of great interest because it is a solution-phase ionization technique applicable to most nonvolatile compounds. In ion evaporation, a volatile
buffer added to the HPLC mobile phase (typically 0.1 M ammonium acetate) can result in statistical charging of the micrometer-size droplets from
the vaporizer (30). These droplets can have a mean surface field strength
of 10^7 volts-per-meter, which increases as the droplets continue to de-
solvate. At sufficiently high surface fields, evaporation of the ions from
the droplet becomes as favorable as evaporation of the neutrals (desolvation). Essentially, ion evaporation is analogous to FD and is well suited
for HPLC because there are no requirements for the compound to be in
the gas phase. In the absence of buffer ions, polar compounds can generate
a field to ionize themselves. Nonpolar molecules require that an auxiliary
ionization method (CI) be employed or a buffer be added postcolumn (31).

The variety of ionization techniques available in thermospray places

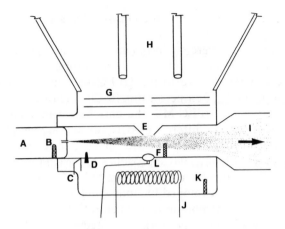

Schematic of Thermospray Interface

A   Direct heated vaporizer
B   Vaporizer thermocouple
C   Filament*
D   Discharge electrode*
E   Ion exit cone
F   Aerosol thermocouple
G   Lenses
H   Quadrupole assembly
I   Liquid nitrogen trap and forepump
J   Source block heater
K   Source thermocouple
L   Repeller

*Dependent on model and manufacture

Figure 7.3. Schematic of a thermospray interface. Reprinted with permission from R. D. Voyksner, J. T. Bursey, E. D. Pellizzari, *Anal. Chem. 56,* 1508. Copyright 1984 American Chemical Society.

few restrictions on the types of samples or HPLC conditions that can be used. Most normal or reverse phase solvents can be used. Buffers necessary for an HPLC separation must be volatile, but buffers necessary only for ion evaporation can be added postcolumn to avoid changing a separation (31). Thermospray typically shows $[M + H]^+$ ion for samples with high proton affinity (PA). Otherwise $[M + NH_4]^+$ (NH_4 is from the ammonium acetate buffer) or $[M + solvent]^+$ ions are detected for samples of lower PA, such as carboxylic acids (32). Sometimes $[M + Na]^+$ is detected in ion evaporation from residual sodium present in the system. Many thermospray interfaces are also equipped with a filament or discharge to obtain gas-phase ionization (CI). CI processes put fewer restrictions on the interface, allowing the use of normal phase solvents and nonbuffered solvents. Also CI processes expand the range of compounds that can be analyzed by the interface. CI of the analyte occurs through gas phase interaction with the HPLC solvent reagent ions (for example, hexane + H^+, H_3O^+, $CH_3OH_2^+$, . . .) to form analyte ions. Because the PA of hexane or water or methanol is lower than ammonia (used in ion evaporation), a wider range of compounds (especially compounds with low PA such as carboxylic acids) can be analyzed. When the filament or discharge is operated in negative ion operation, $[M]^-$ ions are observed if the compound has a high electron-capture cross section. Otherwise, $[M - H]^-$ or $[M + \text{buffer anion}]^-$ ions are detected. Whether $[M - H]^-$ or $[M + \text{buffer anion}]^-$ is observed depends on the gas-phase acidity of the buffer and sample. If trifluoroacetic acid (TFA) is used as a buffer, $[M + TFA]^-$ is the primary ion, whereas ammonium acetate might give primarily $[M - H]^-$ ions (33). The ions discussed thus far are either molecular ions or adducts. Fragmentation can occur but because thermospray makes use of soft ionization techniques, the appearance of fragment ions is difficult to predict.

Thermospray HPLC/MS has been demonstrated on nonvolatile and unstable compounds such as peptides and proteins. For example, thermospray was used to detect glucagon (mw 3483) by observing the $[M + 3H]^{3+}$ ion (34). Thermospray is also very sensitive with on-column detection limits in the low picogram range for many compounds using selected ion monitoring.

Thermospray has several shortcomings. First, day-to-day reproducibility of ionization efficiency is difficult to maintain. Second, fragmentation patterns and sensitivity depend greatly on solvent composition and interface temperature, and in some cases, different temperatures are needed to ionize different compounds efficiently. Although limited fragmentation from thermospray is a drawback for many classes of compounds, the use of tandem MS (MS/MS) can overcome the lack of fragmentation, as is discussed in Section 3.6.

2.4. Mechanical Transport Interface

Another approach to combined HPLC/MS involves solvent removal before the sample is introduced into the MS. Because HPLC solvents are quite volatile relative to most analytes, the solvent can be removed prior to introduction into the MS, eliminating the pumping problems faced in MS with normal HPLC flow rates. Scott et al. (35) recognized this principle and applied it to HPLC/MS by using a moving wire system. The HPLC effluent was deposited on a wire that went through vacuum locks and into the MS source. Although this initial version could accept only

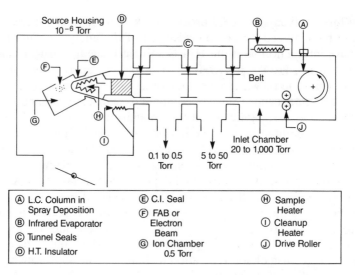

Figure 7.4. Schematic of a moving belt interface.

about 10–50 μL/min in HPLC flow, McFadden et al. (36) optimized the technique to accept normal HPLC flow rates of 1–2 mL/min by replacing the wire with a belt, which, combined with spray deposition (37), enabled the interface to handle larger flow rates.

Currently, moving belt interfaces (Fig. 7.4) spray-deposit the HPLC effluent onto the belt. Spray deposition removes most of the HPLC solvent and maintains chromatographic integrity. The thin solvent–solute film is carried into an optional infrared heating zone through the vacuum locks and into the source. The degree of desolvation is controlled by the power to the vaporizer and infrared heaters. The sample is either flash evaporated from the belt for gas-phase ionization or desorbed from the belt by techniques such as FAB or laser desorption. As the belt exits the MS, a cleanup heater removes residual solvent and sample that could cause ghost peaks.

The moving belt interface is very versatile and sensitive. Samples that are volatile and do not thermally decompose can be analyzed by the gas-phase ionization techniques of EI and CI. Historically, most mass spectral information has been acquired under EI conditions. Comparisons of the EI unknown spectrum to EI spectra of knowns can aid in identifying an unknown. Because CI spectra generally produce molecular ions, CI capabilities allow the molecular weight of the compound to be confirmed or detected (if not observed in EI). The belt interface also can be applied to the analysis of nonvolatile analytes by use of desorption ionization techniques. In desorption ionization, the complete removal of the HPLC solvent, which can be difficult for polar solvents such as water, is unnecessary. The desorption ionization techniques usually result in [M + H]$^+$, [M + Alkali metal]$^+$, [M − H]$^-$, or [M]$^-$ ion formation with limited fragmentation. The recent combination of desorption ionization techniques with the moving belt will prove very valuable to future HPLC/MS work.

The moving belt interface places few restrictions on the HPLC system and can accommodate most HPLC mobile phases and both volatile and nonvolatile buffers. The utility of the interface has been demonstrated in the analysis of a large variety of compounds in complex matrices with picogram sensitivities. However, the interface has several drawbacks including its expense and complexity. Thermally labile compound analysis is somewhat limited and the interface can suffer from contamination when the remaining solvent or sample is not adequately removed before recycling.

3. QUALITATIVE APPLICATION OF HPLC/MS TO CLINICAL TOXICOLOGY

Most HPLC/MS applications have been qualitative where known compounds have been identified in complex mixtures or new compounds were detected and possibly identified. Work applicable to clinical toxicology has been generated primarily from DLI, thermospray, and moving belt interfaces. The choice of which interface is used for a given problem has been based on the compound, instrumental capabilities available, and desired goal. There are few reported comparisons between HPLC/MS techniques to help the potential user choose one interface over another. This section reviews some qualitative applications of HPLC/MS to solve problems in clinical toxicology including the analysis of steroids, alkaloids, mycotoxins, drugs, and pesticides in complex biological and environmental matrices.

3.1. Steroids

Analysis of steroids is difficult because of their nonvolatility, making the use of GC or GC/MS techniques difficult. However, numerous steroids have been analyzed by HPLC/MS. The moving belt interface under CI conditions has been shown to detect betamethasone, dexamethasone, hydrocortisone, prednisolone, cortisone, prednisone, flumethisone, and fluocortolone in horse urine (38). These steroids can be detected down to 50 ng (on column) in urine using negative chemical ionization. The spectra of the steroids contain molecular ions and several fragments. DLI and DLI-micro-HPLC/MS have been demonstrated for the analysis of several steroids including 5α-3-androstanone, estrone, andrastone, betamethasone, cortisone, dexamethasone, corticosterone, diethylstilbestrol, and hydroxyprednisolone (39–42). DLI-HPLC/MS also was reported for the glucuronide conjugate of estriol-17β and 5β-pregnane3α,20β-diol (43). The DLI spectra exhibited $[M - H]^-$ ions and losses of H_2O. Sensitivities of DLI-HPLC/MS employing a 1:100 split are in the microgram range, but the use of DLI-micro-HPLC/MS removed the split enabling detection down to 10–30 ng (on-column, full scan) for dexamethasone and diethylstilbestrol (42).

The use of thermospray HPLC/MS has been demonstrated for the analysis of steroid monosulfates and disulfates (44) (32 different compounds), as well as 18 steroid monoglucuronide conjugates (45). The negative ion spectra of these steroids were acquired under ion evaporation conditions without a buffer. The $[M]^-$ ion was the only ion in the spectra for two glucuronide steroids as shown in Fig. 7.5. Sensitivities of 100 pg were achieved for the steroids and their glucuronide conjugates using a thermospray interface with selected ion monitoring. Steroid conjugates of 1-dehydrotestosterone have been analyzed by FAB/MS (46) and both sulfate and glucuronide conjugates were identified. Detection of the steroids by FAB/MS indicated the potential application of FAB ionization in HPLC/MS using the moving belt interface. Corticosteroids from equine urine at the nanogram-per-milliliter level have recently been determined by SFC/MS (47).

3.2. Alkaloids

Alkaloids are powerful naturally occurring compounds with a variety of therapeutic applications. HPLC/MS using the moving belt interface was used to analyze the ergot alkaloids including agroclavine, festuclavine, setoclavine, palliclavine, penniclavine, isochanoclavine, chanoclavine, noragroclavine, and elymoclavine (48). These alkaloids could be identified in a fermentation batch using both EI and CI techniques. The same workers have analyzed amaryllidaceve and cinchona alkaloids using moving belt HPLC/MS with combined EI and CI (49).

Several alkaloids were analyzed by DLI-HPLC/MS including bromocriptine, reserpine, codergocrine, mesylate, and ergotamien (50–52). The DLI spectra for these alkaloids exhibited an $[M + H]^+$ or $[M]^-$ ion with several fragment ions. It was estimated that 10 ng of bromocriptine could be detected under full scan conditions.

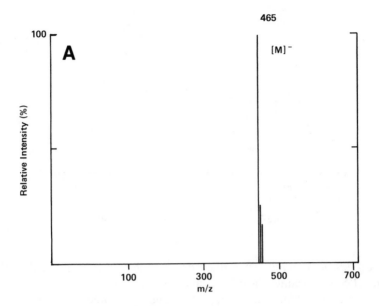

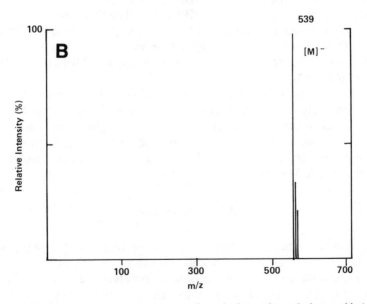

Figure 7.5. Thermospray negative ion spectra of tetrahydrocortisone-3-glucoronide (A) and actiocholan-3α-ol-17-one glucuronide (B). From Ref. (45), with permission of John Wiley & Sons, Ltd.

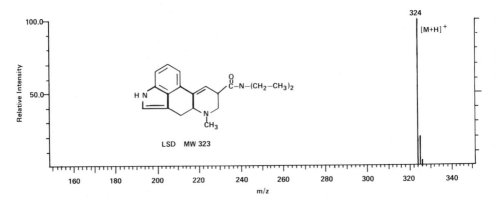

Figure 7.6. Thermospray spectrum of LSD.

Alkaloids also include such illicit drugs such as D-lysergic acid diethylamide (LSD), cocaine, morphine, and heroin. These low-volatility compounds must be identified and their presence validated in biological matrices. HPLC/MS has proved useful for the analysis of these drugs as well as several of their metabolites. LSD has been analyzed by DLI (53) and in our laboratory by thermospray HPLC/MS. The thermospray spectrum (Fig. 7.6) exhibited a single ion, $[M + H]^+$, under ion evaporation conditions. Detection limits using selected ion monitoring were estimated to be 300–500 pg. Comparable results were obtained by DLI-HPLC/MS for LSD in urine (53).

Heroin (54), cocaine (51), morphine, and several of their metabolites have been analyzed by thermospray HPLC/MS. For example, the thermospray/MS spectrum of cocaine and morphine-3-glucuronide are shown in Fig. 7.7. Like LSD, these drugs exhibited an $[M + H]^+$ ion with few fragment ions. Impurities in heroin have also been detected using thermospray HPLC/MS (54).

3.3. Mycotoxins

Mycotoxins have been of interest in our laboratory over the past several years because of their occurrence in grain consumed by animals and the lack of available analytical methodology to detect these toxins directly and quickly. This class of compounds can be analyzed by GC or GC/MS only with derivatization. HPLC with detectors based on photometric detection are subject to matrix interferences. Thermospray HPLC/MS using both positive and negative ion detection has been evaluated for the anal-

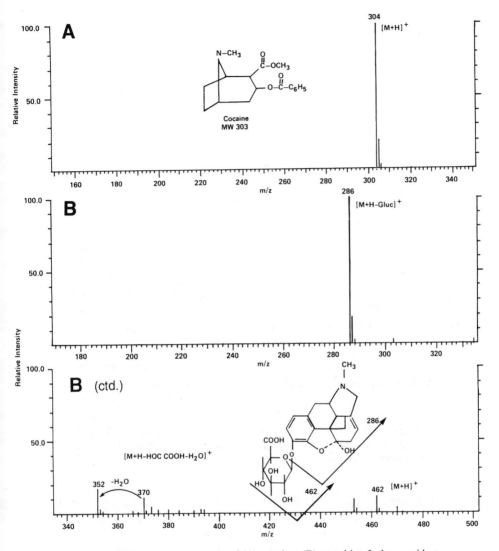

Figure 7.7. Thermospray spectra of (A) cocaine, (B) morphine-3-glucuronide.

ysis of numerous toxins including T-2 toxin, HT-2, diacetoxyscirpenol (DAS), deoxynivalenol (DON), zeralonene, scipentriol, T-2 triol, and nesolanial, as well as their de-epoxy and hydroxy metabolites (55, 56). Operating the thermospray HPLC interface in the CI mode proved to be more sensitive than ion evaporation. The mass spectra of the toxins consisted of $[M + NH_4]^+$ ($[M]^-$ or $[M + acetate]^-$ in negative ion detection)

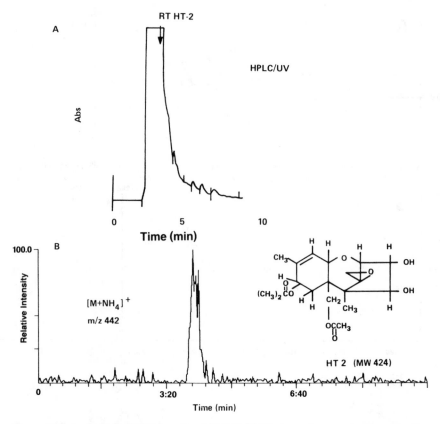

Figure 7.8. Analysis for HT-2 in urine. (A) HPLC/UV chromatogram showing retention time of HT-2. (B) Selected ion chromatogram for the $[M + NH_4]^+$ ion (m/z 442) of HT-2 for 100 pg injected. (Reprinted, with permission, from Ref. (55).

with fragments indicative of the various substitutions on the ring structure of the toxins. Thermospray HPLC/MS in the selected ion monitoring mode was capable of detecting 10–50 pg of most toxins in extracts of plasma, urine, and feces. Thermospray HPLC/MS was not affected by most matrix interferences. For example, compare the chromatogram for the $[M + NH_4]^+$ ion for HT-2 in urine to the HPLC/UV chromatogram for the same sample (Fig. 7.8). Free DOM-1 (de-epoxy metabolite of DON) and its glucuronide conjugate from cow urine, were measured directly by thermospray HPLC/MS (Fig. 7.9). This technique has aided in the determination of a new metabolite of T-2 toxin postulated to be de-epoxy 3′-hydroxy T-2 triol (55, 56).

Techniques including the moving belt, DLI, SFC, and API-HPLC/MS

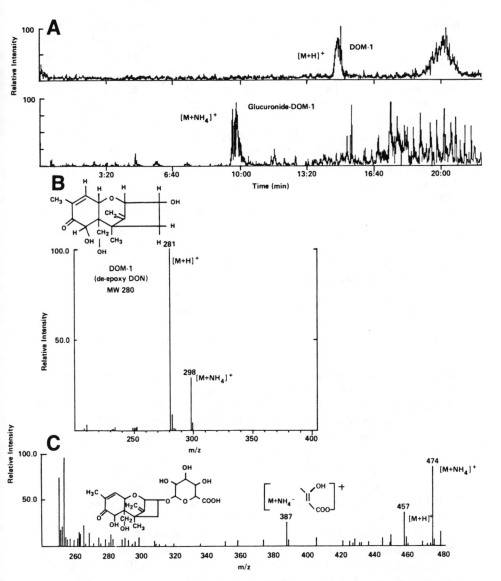

Figure 7.9. HPLC/MS analysis of cow urine for DOM-1 and its glucuronide conjugate. (A) Selected ion chromatograms for the [M + H]⁺ ion of DOM-1 (*m/z* 281) and [M + NH₄]⁺ (*m/z* 474) ion of DOM-1 glucuronide. (B) Thermospray spectrum of DOM-1 (mw 280). (C) Thermospray spectrum of DOM-1 glucuronide (mw 456). With permission from Ref. (56).

have been shown by others to be useful in the analysis of mycotoxins. Detection of 50–100 pg of T-2, HT-2, monoacetoxyscirpenol (MAS), DAS, and T-2 triol has been achieved on-column. Selected ion monitoring using the moving belt interface under ammonia CI conditions and multiple ion detection were used (57). Aflatoxin B_1, B_2, G_1, and G_2 were also analyzed by moving belt interface under methane CI conditions (58). The aflatoxins showed an $[M + H]^+$ ion and could be detected down to 3–30 ng. DON and Novalenial were detected by DLI-HPLC/MS down to the 10-ng range using selected ion monitoring (59). The spectra primarily exhibited $[M + H]^+$ ions with few fragments. The toxins T-2 and DAS were extracted from grains for direct analysis by SFC/MS (60, 61). The SFC/MS spectra showed an $[M + H]^+$ ion and numerous fragment ions for injections of 20 ng of T-2 or DAS. Toxins in grains could be analyzed directly in a few minutes by this technique. Zeralenone and zeranalenol in animal tissue were analyzed down to the parts-per-billion level using API-HPLC/MS (62). The method used a 3-cm column for quick analysis and relied on the specificity of MS/MS to detect the mycotoxins.

3.4. Drugs and Pharmaceuticals

There are numerous reports of the use of HPLC/MS for the analysis of drugs and pharmaceuticals, to aid in metabolite identification, or to analyze target compounds in complex biological matrices. Drugs including 2-hydroxypromazine, acepromazine, chlorpromazine, ephedrine·HCl, procaine·HCl sulfadimethoxine acid, furosemide, ibuprofen, mefenamic acid, dilantin, and meclofenamic acid have been analyzed by Henion and co-workers using DLI-HPLC/MS and DLI micro HPLC/MS (63–65). DLI-HPLC/MS using a chiral column resolved optical isomers of ibuprofen benzylamide, N-1-naphthoylamphetamine, fenoprofen, 1-naphthalenylmethylamide, and benoxaprofen-1-naphthalenylmethylamide (66). Different optical isomers of the drugs show different biological activity. The HPLC separation of these optical isomers was necessary because the DLI-HPLC/MS spectra were identical. Other compounds including cannabis extracted from leaves (67) and several glycosides (43) were analyzed by DLI-HPLC/MS. The technique shows 100–1000-pg (on-column, selected ion monitoring) detection limits with micro HPLC separation; however, the 1:100 split required by the DLI interface with conventional HPLC raises the detection limits to 10–100 ng.

A number of sulfa drugs including sulfamethazine, sulfisoxazole, sulfadiazine, and sulfadimethoxine were analyzed by HPLC/MS and HPLC/MS/MS using an API source (68). The HPLC/MS/MS analysis improved the specificity and provided more structural information other than the simple $[M + H]^+$ ion spectra. Detection limits of 0.5–1 μg were reported for the technique.

Imipramine, promazine, and chloropromazine were analyzed by thermospray HPLC/MS (69). The spectra were nearly identical to those obtained by DLI-HPLC/MS (primarily $[M + H]^+$ ions) with detection limits down to the 1–10 ng range using full-scan detection. Thermospray HPLC/MS was used to study the metabolism of furosemide (70), antibiotics (penicillins and cephalosporins) (71), ampicillin, mezocillin (72), and carnitine (73). Numerous potential anticancer drugs (74, 75) and antimalarial drugs (76) were analyzed by thermospray in a variety of matrices to help identify impurities in the drug formulation. Most of these drugs or impurities were detected (on-column) down to the 10–200 ng level for full scan and 1–5 ng using selected ion monitoring.

The moving belt interfaced HPLC/MS was used in the analysis of disopyramide (77) as well as to study the metabolism of ranitidine (78).

3.5. Pesticides

There are numerous references to the use of HPLC/MS for the analysis of pesticides and herbicides in environmental matrices. Some major classes of pesticides including carbamates, triazines, organophosphates, and chlorinated phenolic acids have been analyzed by the moving belt (79–81), DLI (82–86), and thermospray (31, 32, 69, 87–89) HPLC/MS. The spectra obtained from these three interfaces can be compared for the analysis of propoxur (Fig. 7.10). The moving belt interface can operate under EI and CI conditions resulting in greater fragmentation and more structural information in the spectra of propoxur. Thermospray and DLI-HPLC/MS, on the other hand, showed only $[M + H]^+$ ions for pesticides. At times, the lack of fragmentation with DLI and thermospray interfaces was overcome with the use of HPLC/MS/MS (90). Detection limits for the analysis of pesticides in the 5–50 pg levels were achieved with the belt and thermospray interfaces, whereas DLI exhibited detection limits closer to the 5–10 ng range because of the solvent split. The HPLC/MS technique proved very specific for the analysis of pesticides in complex matrices.

Aldicarb residues in water (91) were examined using HPLC with the moving belt interface and several aldicarb residues were identified. The same technique was used to study the metabolism of chloropropham in rats and new metabolites of chloropropham were identified in urine (42, 92).

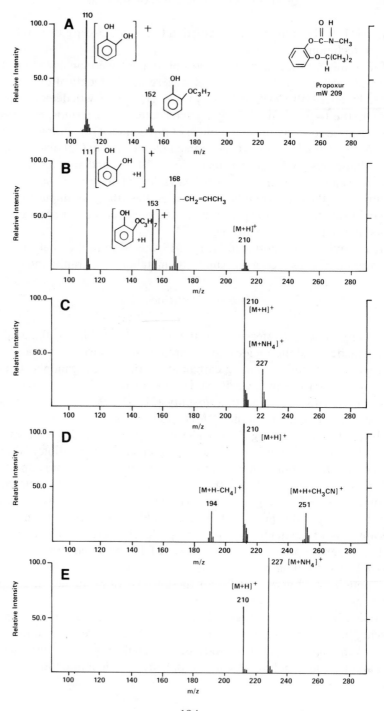

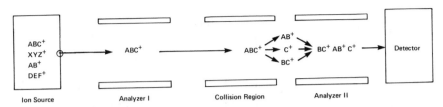

Figure 7.11. Principle of MS/MS analysis.

3.6. HPLC/MS/MS

The use of MS/MS with HPLC has become increasingly important in recent years. MS/MS offers the capability of added structural information for compounds producing spectra containing a single or a few ions. MS/MS also adds extra specificity, reduction of chemical noise, and the ability to remove HPLC solvent background ions to detect ions in the lower mass range ($m/z \leq 150$). The basics of MS/MS have been well reviewed (93–95) and numerous reports of its use are found in the literature. The basic concept of the technique (illustrated in Fig. 7.11) is that the first analyzer selects a desired m/z ion from all the ions in the source. This ion is collisionally activated with a neutral gas to cause the ion to fragment. The fragment ions formed are mass analyzed by the second analyzer and detected (daughter ion spectrum). The MS/MS spectrum of a molecular ion can resemble an EI spectrum for the compound. Background from the solvent and matrix with different masses are removed by the first analyzer, greatly reducing chemical noise. Compounds that co-elute in HPLC can often be separated by MS/MS. However, isobaric or isomeric compounds, which are often separated by HPLC, can be resolved by MS/MS only if the daughter ions differ. For these reasons, the combination of HPLC/MS/MS can provide an extra dimension of specificity not achieved in the individual techniques. The capabilities of MS/MS have been demonstrated in the analysis of classes of compounds including steroids (71), alkaloids (48), mycotoxins (57, 61, 62), drugs (68, 71), and pesticides (90). In all cases more structural information and reduction in chemical noise were gained in the analysis.

An example of the capabilities of HPLC/MS/MS is shown for the anal-

---

Figure 7.10. HPLC/MS spectra of propoxur obtained from moving belt interface, using (A) EI conditions, (B) methane CI, and (C) ammonia CI; (D) DLI interface; and (E) thermospray interface. Moving belt data (B and C) reprinted with permission from J. Stamp, E. Siegmund, T. Cairrs, K. Chan, *Anal. Chem. 58*, 873. Copyright 1986 American Chemical Society. (A) from ref. (102) with permission of L. Slivon of Battelle, Columbus Labs, Columbus, Ohio.

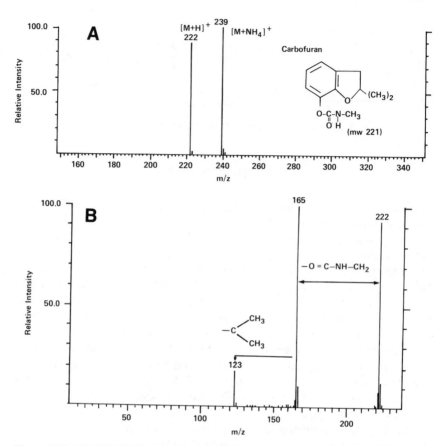

Figure 7.12. (A) HPLC/MS spectrum of carbofuran, (B) HPLC/MS/MS spectrum of the [M + H]$^+$ ion (m/z 222) of carbofuran. Both spectra were acquired by thermospray under ion evaporation conditions using the same HPLC system and analysis conditions.

ysis of the pesticide carbofuran. The DLI or thermospray HPLC/MS spectra for this pesticide consisted of only an [M + H]$^+$ and [M + NH$_4$]$^+$ ion at m/z 222 and m/z 239, respectively (Fig. 7.12A). However, in HPLC/MS/MS, m/z 222 is selected by the first analyzer, collisionally activated, and the fragment ions formed from the activated ion are detected through the second analyzer to produce a daughter ion spectrum shown in Fig. 7.12B. The fragment ions detected are from simple cleavages of the alkyl side groups on the molecule. These fragment ions are necessary to confirm the presence of this pesticide rather than some other compound that has the same molecular weight as carbofuran. This is important in an envi-

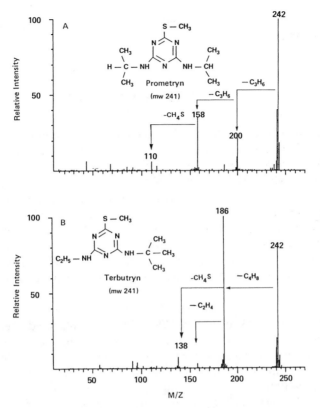

Figure 7.13. HPLC/MS/MS spectra of two isomeric triazines (A) prometryn and (B) ter-butryn (mw 241).

ronmental sample because one ion is not sufficient to confirm an identification.

Another example (Fig. 7.13) shows how HPLC/MS/MS can be used to identify two isomeric triazines in a mixture. The triazines, which are resolved by HPLC, cannot be identified unless there are reference standards for each. The MS/MS spectrum of the ion at m/z 242 is significantly different for the compounds, which aids in the postulation of the two structures, shown in Fig. 7.13. With this information, reference standards can be purchased or synthesized to verify the postulated identity.

The increased information available from HPLC/MS/MS combined with decreasing instrument cost promise increased applications of this technique in the future. The ability of this technique possibly to determine unknown compounds in complex matrices and metabolites with no standards makes it very desirable and will promote growth in the field of HPLC/MS/MS.

4. QUANTITATIVE HPLC/MS

Quantitative HPLC/MS, unlike the qualitative applications, is still in its infancy. Only a limited amount of work has been reported that uses quantitative HPLC/MS, primarily for the analysis of drugs in biological fluids. Quantitative HPLC/MS has probably received little attention because analyses using the various interfaces for HPLC/MS are much less reproducible than by GC/MS. The effect of instrumental instability, which usually limits the accuracy of quantitation, can be reduced by using isotopically labeled internal standards. However, compounds that require HPLC are more complex than those that can be analyzed by GC and therefore are difficult and expensive to synthesize as isotopically labeled analogs.

Isotopically labeled internal standards have been used in several quantitative determinations using HPLC/MS. Deuterium-labeled internal standards were used to analyze a bronchodilator (96), choline and acetylcholine (97), 9-keto cannabinoid (98), and carnitine and its acyl derivatives (73) by thermospray HPLC/MS. The metabolism of ranitidine (78) and disopyramide (99) was studied using the moving belt interface. Ion evaporation thermospray HPLC/MS was used to determine ^{15}N enrichment of NO_3^- and NO_2^- in plants and soil from nitrogen-cycling bacteria (100). Standard curves generated for the analysis in the above examples were linear over the range of interest (usually two to three orders in magnitude in concentration), with detection in the nanogram to microgram level range in real biological matrices.

There are several reports of quantitation by HPLC/MS using a homolog internal standard or by external calibration only. Hydrochlorothiazide was determined in equine urine by DLI-HPLC/MS down to 25 ng (101). Disopyramide, using a chloro analog as an internal standard, was quantitated down to the 1–10 ng level (99). The analysis of the carbamate pesticide chlorpropham by the moving belt interface showed a linear response over the range of 1 ng to 10 μg (79). The analysis of rotenone by thermospray HPLC/MS proved linear from 1 ng to 1 μg (89).

Clearly these few examples demonstrate that HPLC/MS can be applied to quantitative problems. The quantitative application of HPLC/MS is expected to increase in the near future because of interfaces with greater reproducibility and increased HPLC/MS sensitivities.

REFERENCES

1. J. T. Watson, *Introduction to Mass Spectrometry*, 2nd ed., Raven Press, New York, 1985.
2. M. Barber, R. Brodolli, G. Elliott, D. Sedgwick, and A. Tyler, *Anal. Chem.*, *54*, 645A (1982).
3. H. Schulten, *J. Chromatogr.*, *251*, 105 (1982).
4. M. Posthumus, P. Kistemaker, H. Meuzelaar, and M. de Braun, *Anal. Chem.*, *50*, 985 (1978).
5. M. Vestal, *Science*, *226*, 275 (1984).
6. W. McFadden, *J. Chromatogr. Sci.*, *17*, 1 (1979).
7. G. Guiochan and P. Arpino, *Anal. Chem.*, *51*, 632A (1979).
8. D. Desiderio and G. Fridland, *J. Liq. Chromatogr.*, *7*, 317 (1984).
9. N. Nibbering, *J. Chromatogr.*, *251*, 93 (1982).
10. D. Games, N. Alcock, I. Horman, E. Lewis, M. McDowall, and A. Moncur, *Chromatography and Mass Spectrometry in Nutrition Science and Food Safety*, Analytical Chemistry, Symposium Series, Vol. 21, Elsevier, Amsterdam, 1984.
11. J. Huber, A. Van Urk-Schroen, and G. Sieswerda, *Z. Anal. Chem.*, *264*, 257 (1973).
12. V. Tal'Roze, G. Karpov, I. Gordoetshii, and V. Skurat, *Russ. J. Phys. Chem.*, *42*, 1658 (1968).
13. M. Baldwin and F. McLafferty, *Org. Mass Spectrom.*, *7*, 1111 (1973).
14. F. McLafferty, R. Knutti, R. Venkataraghavan, P. Arpino, and B. Dawkins, *Anal. Chem.*, *47*, 1503 (1975).
15. F. Sugnaux, D. Skrabalak, and J. Henion, *J. Chromatogra.*, *264*, 351 (1983).
16. J. Henion, *Anal. Chem.*, *50*, 1687 (1980).
17. M. Dedieu, C. Juin, P. Arpino, J. Bounine, and G. Guiochon, *J. Chromatogr.*, *251*, 203 (1982).
18. A. Melera, *Adv. Mass Spectrom.*, *8b*, 1597 (1980).
19. M. Novotny, S. Springston, P. Peaden, J. Fjeldsted, and M. Lee, *Anal. Chem.*, *53*, 407A (1981).
20. R. Smith, W. Felix, J. Fjeldsted, and M. Lee, *Anal. Chem.*, *54*, 1883 (1982).
21. R. Smith, J. Fjeldsted, and M. Lee, *Int. J. Mass Spectrom. Ion Phys.*, *46*, 219 (1983).
22. D. Carroll, I. Dzidic, R. Stillwell, K. Haegele, and E. Horning, *Anal. Chem.*, *47*, 2369 (1975).
23. T. Covey, E. Lee, A. Bruins, and J. Henion, *Anal. Chem.*, *58*, 1451A (1986).
24. B. Thomson, I. Iribarne, and P. Dziedzic, *Anal. Chem.*, *54*, 2219 (1982).
25. P. Jones and S. Yang, *Anal. Chem.*, *47*, 1000 (1975).
26. C. Blakely, J. Carmody, and M. Vestal, *J. Am. Chem. Soc.*, *102*, 5931 (1980).

27. C. Blakely, M. McAdams, and M. Vestal, *J. Chromatogr., 158,* 261 (1978).

28. C. Blakely, J. Carmody, and M. Vestal, *Anal. Chem., 52,* 1636 (1980).

29. C. Blakely and M. Vestal, *Anal. Chem., 55,* 750 (1983).

30. E. Dodd, *J. Appl. Phys., 24,* 73 (1953).

31. R. Voyksner, J. Bursey, and E. Pellizzari, *Anal. Chem., 56,* 1507 (1984).

32. R. Voyksner and C. Haney, *Anal. Chem., 57,* 991 (1985).

33. C. Parker, R. Smith, S. Gaskell, and M. Bursey, *Anal. Chem., 58,* 1661 (1986).

34. D. Pilosof, H. Kim, K. Dyckes, and M. Vestal, *Anal. Chem., 56,* 1236 (1984).

35. R. Scott, C. Scott, M. Munroe, and J. Hess, Jr., *J. Chromatogr., 99,* 395 (1974).

36. W. McFadden, H. Schwartz, and S. Evans, *J. Chromatogr., 122,* 389 (1976).

37. E. Lankmayr, M. Hayes, B. Karger, P. Vouros, and J. McQuire, *Int. J. Mass Spectrum Ion Phys., 46,* 177 (1983).

38. E. Houghton, M. Dumasia, and J. Wellby, *Biomed. Mass Spectrom., 8,* 558 (1981).

39. C. Eckers, J. Henion, G. Maylin, D. Skrabalak, J. Vessman, A. Tivert, and J. Greenfield, *Int. J. Mass Spectrom. Ion Phys., 46,* 205 (1983).

40. P. Arpino, B. Dawkins, and F. McLafferty, *J. Chromatogr. Sci., 12,* 574 (1974).

41. P. Arpino, M. Baldwin, and F. McLafferty, *Biomed. Mass Spectrom., 1,* 80 (1974).

42. J. Henion and D. Wachs, *Anal. Chem., 53,* 1963 (1981).

43. C. Kenyon, P. Goodley, D. Dixon, J. Whitney, K. Faull, and J. Barchas, *Am. Lab., 16,* 38 (1983).

44. D. Watson, G. Taylor, and S. Murray, *Biomed. Environ. Mass Spectrom., 12,* 610 (1985).

45. D. Watson, G. Taylor, and S. Murray, *Biomed. Environ. Mass Spectrom., 13,* 65 (1986).

46. M. Dumasia, E. Houghton, C. Bradley, and D. Williams, *Biomed. Mass Spectrom., 10,* 434 (1983).

47. E. Lee and J. Henion, 34th Annual Conf. Mass Spectrom. Allied Topics, Cincinnati, OH, Paper FPA13, 1986.

48. C. Eckers, D. Games, D. Mallen, and B. Swann, *Biomed. Mass Spectrom., 9,* 162 (1982).

49. C. Eckers, D. Games, E. Lewis, K. Nagaraja Rao, M. Rossiter, and N. Weerasinghe, *Adv. Mass Spectrom., 8,* 1396 (1980).

50. C. Kenyon, A. Melera, and F. Erni, *J. Anal. Toxicol., 5,* 216 (1981).

51. F. Erni, *J. Chromatogr., 251,* 141 (1982).

52. J. Henion, *J. Chromatogr. Sci., 18,* 101 (1980).

53. C. Kenyon and A. Melera, *J. Chromatogr. Sci., 18,* 103 (1980).

54. J. Joyce, R. Andrey, and I. Lewis, *Biomed. Mass Spectrom.*, *12*, 588 (1985).

55. R. Voyksner, W. Hagler, K. Tyczkowska, and C. Haney, *J. High Resol. Chromatogr. Chromatogr Comm.*, *8*, 119 (1985).

56. R. Voyksner, W. Hagler, and S. Swanson, *J. Chromatogr.*, *394*, 183 (1987).

57. S. Missler, 33rd Annual Conf. Mass Spectrom. Allied Topics, San Diego, CA, 659, 1985.

58. W. McFadden, D. Bradford, D. Games, and J. Gower, *Am. Lab*, *9*, 55 (1977).

59. R. Tiebach, W. Blaas, M. Kellert, S. Stienmeyer, and R. Weber, *J. Chromatogr.*, *318*, 103 (1985).

60. R. Smith and H. Udseth, *Biomed. Mass Spectrom.*, *10*, 577 (1983).

61. R. Smith and H. Udseth, *Anal. Chem.*, *55*, 2266 (1983).

62. T. Covey and J. Henion, 33rd Annual Conf. Mass Spectrom Allied Topics, San Diego, CA, 575, 1985.

63. J. Henion and G. Maylin, *Biomed. Mass Spectrom.*, *7*, 115 (1980).

64. J. Henion, *Anal. Chem.*, *50*, 1687 (1978).

65. J. Henion, *Adv. Mass Spectrom.*, *8B*, 1241 (1980).

66. J. Crowther, T. Covey, E. Dewey, and J. Henion, *Anal. Chem.*, *56*, 2921 (1984).

67. P. Arpino and P. Krien, *J. Chromatogr. Sci.*, *18*, 104 (1980).

68. J. Henion, B. Thomson, and P. Dawson, *Anal. Chem.*, *54*, 451 (1982).

69. T. Covey, J. Crowther, E. Dewey, J. Henion, *Anal. Chem.*, *57*, 474 (1985).

70. D. Liberato, N. Estebam, A. Rachmel, G. Yost, and A. Yergey, 33rd Annual Conf. Mass Spectrom Allied Topics, San Diego, CA, 747, 1985.

71. S. Unger and B. Warrack, *Spectroscopy*, *1*(3), 33 (1986).

72. C. Blakely, J. Carmody, and M. Vestal, *Clin. Chem.*, *26*, 1467 (1980).

73. A. Yergey, D. Liberato, and D. Millington, *Anal. Biochem.*, *139*, 278 (1984).

74. R. Voyksner, J. Bursey, and J. Hines, *J. Chromatogr.*, *323*, 383 (1985).

75. R. Voyksner, F. Williams, and J. Hines, *J. Chromatogr.*, *347*, 137 (1985).

76. R. Voyksner, J. Bursey, J. Hines, and E. Pellizzari, *Biomed. Mass Spectrom.*, *11*, 616 (1984).

77. D. Games, P. Hirten, W. Kuhnz, E. Lewis, N. Weerasinghe, and S. Westwood, *J. Chromatogr.*, *203*, 131 (1981).

78. L. Martin, J. Oxford, and R. Tanner, *J. Chromatogr.*, *251*, 215 (1982).

79. L. Wright, *J. Chromatogr. Sci.*, *20*, 1 (1982).

80. T. Cairns, E. Siegmund, and G. Doose, *Biomed. Mass Spectrom.*, *10*, 24 (1983).

81. J. Stamp, E. Siegmund, T. Cairns, and K. Chan, *Anal. Chem.*, *58*, 873 (1986).

82. L. Shalaby, *Biomed. Mass Spectrom.*, *12*, 261 (1985).

83. C. Parker, C. Haney, D. Harvan, and J. Hass, *J. Chromatogr.*, *242*, 77 (1982).

84. C. Parker, C. Haney, and J. Hass, *J. Chromatogr., 237,* 233 (1982).

85. R. Voyksner, J. Bursey, and E. Pellizzari, *J. Chromatogr., 312,* 221 (1984).

86. R. Voyksner and J. Bursey, *Anal. Chem., 56,* 1582 (1984).

87. L. Shalaby, 33rd Ann. Conf. Mass Spectrom Allied Topics, San Diego, CA, 773, 1985.

88. L. Shalaby, 33rd Annual Conf. Mass Spectrom. Allied Topics, San Diego, CA, 665, 1985.

89. M. McAdams and M. Vestal, *J. Chromatogr. Sci., 18,* 110 (1980).

90. R. Voyksner, W. McFadden, and S. Lammert, in J. Rosen, Ed., *Applications of New Mass Spectrometry Technologies in Pesticide Chemistry,* Wiley, New York, 1987, p. 247.

91. L. Wright, M. Jackson, and R. Lewis, *Bull. Environ. Contam. Toxicol., 28,* 740 (1982).

92. D. Games, *Biomed. Mass Spectrom., 8,* 459 (1981).

93. R. Kondrat and R. Cooks, *Anal. Chem., 50,* 81A (1978).

94. R. Yost and C. Enke, *Anal. Chem., 51,* 1251A (1979).

95. R. Cooks, *Anal. Chem., 57,* 823A (1985).

96. D. Satonin and J. Coutant, 34th Annual Conf. Mass Spectrom. Allied Topics, Cincinnati, OH, RPB11, 1986.

97. D. Liberato, A. Yergey, and S. Weintraub, *Biomed. Environ. Mass Spectrom., 13,* 171 (1986).

98. H. Sullivan and D. Garteiz, 33rd Annual Conf. Mass Spectrom. Allied Topics, San Diego, CA, 198, 1981.

99. D. Games, E. Lewis, N. Haskins, and K. Waddell, *Adv. Mass Spectrom., 8,* 1233 (1981).

100. L. Hogge, M. Vestal, R. Hynes, L. Nelson, and S. Sask, 33rd Annual Conf. Mass Spectrom. Allied Topics, San Diego, CA, 629, 1985.

101. J. Henion and G. Maylin, *J. Anal. Toxicol., 4,* 185 (1980).

102. L. Slivon and F. DeRoos, *Development of a General Purpose LC/MS Method for Compounds of Environmental Interest,* EPA Final Report submitted to EPA, Cincinnati, OH, March 3, 1983, by Battelle Columbus Labs.

CHAPTER

8

DRY REAGENT CHEMISTRIES

KENNETH EMANCIPATOR AND DALE G. DEUTSCH

State University of New York at Stony Brook
Stony Brook, New York

1. INTRODUCTION

Clinical toxicology can roughly be divided into two major disciplines: therapeutic drug monitoring (TDM) and toxicological screening. The former is a much simpler task because it involves assaying for a single known drug, whereas the latter involves screening for a large number of chemically dissimilar drugs that are capable of producing poisoning or overdose. Furthermore, TDM is required more frequently than is toxicological screening. Thus, for economic reasons, it is not surprising that the major advances in dry reagent chemistries relevant to the toxicologist are in the area of TDM, although drug abuse screening is growing rapidly.

TDM is required whenever a patient is on one of a number of very commonly used drugs. For each of these drugs, the serum concentration produced by a given dose varies greatly with each individual patient, yet the serum concentration required for a therapeutic effect is relatively close to the concentration that produces a toxic effect. Therefore, it is necessary to titrate the dose to the proper serum concentration. Advances in the techniques of immunoassay have made TDM more widely available. As TDM moved from reference laboratories into community hospitals, there was a great reduction in the turnaround time for these assays. The goal of dry reagent TDM is to reduce turnaround time even further by providing a practical method of obtaining timely results not only in hospital laboratories, but also in physicians' offices, patient wards, or emergency rooms ("bedside testing").

The demand for immediate results is no doubt fostered by the success of capillary blood glucose monitoring in diabetic patients (1–3). Development of dry reagent test strips that are read using handheld reflectance photometers for quantitative glucose measurements has made it possible for patients themselves routinely to adjust insulin doses based on their current blood glucose. This has enabled a degree of control of blood glucose that would otherwise be impossible. Although this level of monitoring has no role in TDM, the idea of having immediate results has an obvious appeal to physicians. There is also an economic advantage to a physician in private practice who performs tests in his office, rather than sending them out to laboratories.

However, there are hazards inherent in bedside testing. When tests are performed outside the laboratory, the responsibility for performing the tests generally falls on personnel not thoroughly trained in laboratory procedures and techniques, that is, physicians, nurses, ward clerks, and receptionists. In these situations, centrifugation of sample, dilution of serum, application of precise amounts of diluted sample to test strips, handling of wet reagents, and precise timing of reflectance measurements with respect to application of sample to test strip are significant disadvantages to an assay. The need for adequate training for individuals who are to perform bedside tests cannot be overemphasized (4). In addition, development of a quality-control program is mandatory.

In cases of suspected overdose or poisoning, laboratory results must be available within a few hours to be useful. Yet for most hospitals, high performance liquid chromatography (HPLC) available on a 24-hr basis is not practical, and transport to reference laboratories adds considerable delay to obtaining the results. Therefore, there is a need for rapid, simple methods for screening for drugs of abuse. Unfortunately, for reasons already mentioned, there has been little progress in this area thus far.

Because drug screening is increasingly in demand in nonclinical situations, for example, as part of a condition for employment, financial incentives for developing rapid, inexpensive toxicological screening procedures are evolving. The techniques that have been developed for TDM will undoubtedly be adapted to toxicologic screening.

2. SUBSTRATE-LABELED FLUORESCENT IMMUNOASSAY

Substrate-labeled fluorescent immunoassay (SLFIA) was originally developed as a homogeneous liquid-phase assay for theophylline (5). It was subsequently adapted to the dry reagent format (6). The SLFIA using a dry reagent test strip has now been described for a wide variety of therapeutic drugs including theophylline, phenytoin, primidone, carbamazepine, gentamicin, tobramycin, and amikacin (7). The assay requires a fluorometer, but all reagents can be incorporated onto a single dry reagent test strip, and the assay can be performed in less than 5 min. Unfortunately, the SLFIA using dry reagents is not yet commercially available.

2.1. Principle of the Assay

There are essentially three reagents present on the test strip: antibody to the drug of interest, drug conjugated to β-galactosyl umbelliferone (GU–drug), and β-galactosidase. The reactions that occur are shown in Fig. 8.1. Drug in the sample competes with GU–drug for binding sites to antibody; as the concentration of drug in the sample increases, the amount of GU–drug bound to antibody decreases, and therefore the amount of unbound GU–drug increases. In the presence of β-galactosidase, the galactosyl residue is hydrolysed from the unbound GU–drug; however, the enzyme is unable to catalyze hydrolysis of GU–drug bound to antibody. The product of hydrolysis, umbelliferone–drug, is fluorescent. The excitation peak occurs at an approximate wavelength of 400 nm and the emission peak occurs at approximately 450 nm. The concentration of drug in the sample is determined from the intensity of the emission; the intensity of emission increases with increasing concentration of drug.

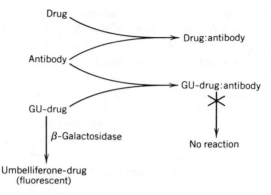

Figure 8.1. Reactions that occur in the substrate-labeled fluorescent immunoassay (SLFIA).

2.2. Preparation of the Test Strip

Chromatography paper is cut into 1×0.5 cm test strips, which are first saturated by an aqueous solution containing antibody to the drug of interest and β-galactosidase, buffered at pH 8.3. The strips are then dried at 50°C. Next, a solution containing GU–drug in an organic solvent such as toluene or acetone is applied to the test strip, which is then dried again at 50°C. Hydrolysis of GU–drug in the presence of galactosidase and binding of GU–drug to antibody do not occur in organic solvent.

2.3. Procedure and Instrumentation

Because the dry reagent SLFIA is not yet commercially available, the procedure described here anticipates that the time required for the reactions on the reagent strip to reach an end point will be reduced to 1 min. The developers of the assay believe this goal is technically feasible (6).

First, a 1:20 dilution of serum is prepared; 35 μL of diluted serum is applied to a 1×0.5 cm test strip. One minute after sample is applied to the test strip, fluorescence is measured. Concentration is determined from fluorescence by comparison to a linear standard curve. The standard curve is determined by a two-point calibration method.

A detailed description of the design of the reflectance fluorometer is given elsewhere (8). Basically, the test strip is illuminated by light from a 40-W mercury lamp passed through a 405-nm three-cavity interference filter. The angle of incidence of the illuminating light is 30° from the axis normal to the plane of the test strip. The light emitted at an angle of 60°

from the normal axis is passed through a 450-nm three-cavity interference filter, and the intensity is measured by a photomultiplier tube.

2.4. Applications

The SLFIA has several distinct advantages. First, at low concentrations of drug, fluorescence will vary linearly with drug concentration (Fig. 8.2). This linear relationship not only makes two-point calibration feasible; it also obviates the need for elaborate mathematical transformations that must be employed in reflectance photometry. Second, the SLFIA is an end point assay, so the timing of the fluorescence measurement relative

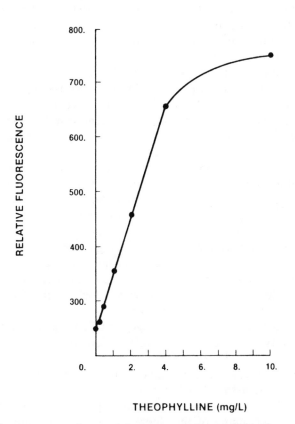

Figure 8.2. Fluorescence vs. the actual drug concentration applied to the reagent strip for the substrate-labeled fluorescent immunoassay for theophylline. Reprinted with permission from *Clinical Chemistry*, 27(9), 1617 (1981). Copyright 1981 The American Association for Clinical Chemistry.

to the application of the diluted sample to the test strip is not critical. Third, fluorescence measurements are more sensitive and are applicable over a wider dynamic range than are reflectance measurements (6). Finally, all reagents necessary for the assay can be incorporated into a single dry reagent test strip.

The disadvantages of the SLFIA are that the assay must be performed on diluted serum and that the instrument required to perform the assay is slightly more complex than a reflectance photometer. Dilution of the sample is necessary so that the actual concentration of drug applied to the test strip is low enough so that it falls on the linear portion of the dose–response curve (Fig. 8.2).

A note of caution is in order here. In the preceding section, it was anticipated that the end point of the assay could be reached within 1 min. In actual experiments, the end point was not reached until approximately 4 min after application of sample to the test strip. In order to reduce the time required for completion of the assay, fluorescence measurements were made before end point was reached, that is, at 3 min after sample was applied to the test strip (6, 7). An untested assumption is that the time required to reach end point can easily be reduced to 1 min by increasing the amount of enzyme on the reagent strip (6). Many other aspects of the assay are also unknown, for example, the stability of the reagent strips during long-term storage.

The advantages afforded by the dry reagent SLFIA over other methods utilizing reflectance photometry would seem to warrant further development. Because the unique advantages of the SLFIA offer the possibility of producing more precise quantitative results than any of the other dry reagent methods, the SLFIA is probably suitable for use in hospital laboratories, yet the procedure is simple enough to be performed in physicians' offices or in "bedside" satellite hospital laboratories.

In a trial in which theophylline assays were performed on clinical samples (7), both the within-run and among-run coefficients of variation were less than 9% at a concentration of 5 mg/L, and less than 4% at concentrations of 15–25 mg/L. The dry reagent SLFIA was found to correlate well with assay by HPLC; the correlation coefficient was 0.98 and the standard error was 1.72 mg/L.

Using this methodology, instrumentation could be tailored to the intended use of the assay. For low-volume testing, dilution of serum, application of diluted serum to the test strip, placement of the strip in the fluorometer, and timing of the fluorescence measurement could all be performed manually. At the other extreme, for very-high-volume testing, all of these steps could be fully automated. The instrument could be designed as a random access TDM analyzer in which samples could be put on the machine in any order, and the appropriate drug assay(s) would be chosen for each sample.

3. APOENZYME REACTIVATION IMMUNOASSAY SYSTEM

The apoenzyme reactivation immunoassay system (ARIS) was first described as a homogeneous liquid-phase assay (9), and shortly thereafter it was adapted to a dry reagent format (10–12). The ARIS is commercially available in the dry reagent format for theophylline, phenytoin, phenobarbital, and carbamazepine. The system has also been used in a thyroxin-binding globulin assay (13) using wet reagents. The ARIS appears to be quite versatile, and successful application to other drug assays can be anticipated. The ARIS offers several major advantages. All reagents necessary for the assay are present on the test strip; that is, no "developing solutions" are required. The only instrument needed is a reflectance photometer. Diluted serum or plasma is simply applied to the test strip and the color change is monitored by the reflectance photometer. The entire test is performed in less than 2 min.

3.1. Principle of the Assay

There are four major components to the ARIS: (1) drug conjugated to flavine adenine dinucleotide (drug–FAD), (2) an antibody to the drug, (3) apoglucose oxidase, the inactive apoenzyme remaining when FAD is removed from glucose oxidase, and (4) reagents that produce a color change in the presence of reactivated glucose oxidase.

The reactions that occur in the assay are shown in Fig. 8.3. The drug–FAD conjugate can combine with apoglucose oxidase to form an active enzyme complex, that is, glucose oxidase. However, when antibody is bound to the drug–FAD conjugate, it cannot reactivate apoglucose oxidase. Drug present in the patient's serum will bind to the antibody, and therefore less antibody binds to the drug–FAD on the test strip. The drug–FAD not bound by antibody reactivates the apoenzyme, which allows the oxidation of glucose, producing hydrogen peroxide. In the presence of peroxidase and hydrogen peroxide, tetramethylbenzidine (TMB), the indicator, is oxidized, producing a blue color. The change in reflectance is measured for quantitative results.

3.2. Preparation of the Test Strip

The reagent paper is prepared in two steps. First, filter paper is immersed in an aqueous solution containing antibody to the drug, apoglucose oxi-

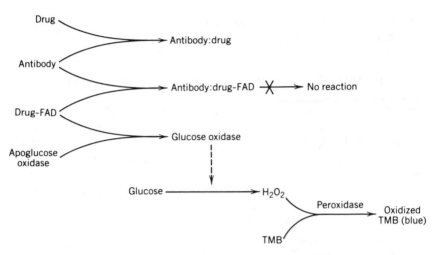

Figure 8.3. Reactions that occur in the apoenzyme reactivation immunoassay system (ARIS).

dase bound to antiglucose oxidase antibody, peroxidase, glucose, and buffer. Antiglucose oxidase antibody is included because it enhances the reactivation of the apoenzyme (14). When FAD is removed from glucose oxidase, there is a conformational change in the protein. This new conformation renders the protein more susceptible to denaturation than the intact enzyme. Furthermore, even if denaturation does not occur, reactivation of the enzyme by FAD is retarded by the need for the protein to return to its active conformation. It has been shown that reactivation does not occur at temperatures above 30°C. Antiglucose oxidase stabilizes the conformation of the apoenzyme, enhancing reactivation, and allowing reactivation at 37°C.

In the second step, the paper is immersed in organic solvent containing drug–FAD and 3,3′,5,5′-tetramethylbenzidine (TMB). The organic solvent prevents binding of drug–FAD to either apoglucose oxidase or antidrug antibody during preparation of the test strip. The test strips are dried in warm air. Induction of the reaction sequence is now prevented by absence of a liquid phase. Thus the test strips must be kept free of moisture during storage.

3.3. Instrumentation and Data Handling

Quantitative results are obtained by measuring the amount of light, at a specific wavelength, which is reflected from the test strip. There are two types of reflection, specular and diffuse. A mirror is an example of specular reflection in which the angle of incidence equals the angle of reflection. Specular reflection occurs because of differences in transmission of light through different media. Light reflected from any dull surface is diffuse; light is reflected in all directions. Diffuse reflection occurs because of collisions between the incident light and the particles of matter. In reflectance photometry, it is the diffuse reflectance that is useful, because the intensity of the diffusely reflected light of a specific wavelength is affected by the type of molecules in the medium.

The measurement of reflectance is quite straightforward. The test strip is uniformly illuminated with diffuse light, and the intensities of the incident illumination and the reflected light are measured at an appropriate wavelength. The reflectance is given by $R = J/I$, where J is the intensity of light reflected from the test strip, and I is the intensity of the source illumination. Note that R ranges from 0 to 1.

A complete description of one instrument designed to make rapid reflectance measurements on small test strips has been given (15). In order to illuminate the test strip with diffuse light, an "integrating sphere" (16) is utilized. This is a spherical cavity, the surface of which is coated with a high reflectance white paint. Figure 8.4 shows the configuration of the ports leading into this cavity. Monochromatic light enters the sphere through a port at one end of the x axis. A reference signal is obtained from light exiting through a port at one end of the y axis, which is measured by a phototube. A port at one end of the z axis allows illumination of the test strip. Light reflected from the test strip exits through a port at the opposite end of the z axis, and is measured by a second phototube.

The optimal wavelength for the detection of oxidized TMB is 660 nm (10). Reflectance at 740 nm was actually measured in later reports (11, 12), because this was the optimal operating wavelength of the instrument that was employed.

Unfortunately, the relationship between concentration of a compound and reflectance does not follow Beer's law. Therefore, reflectance is converted to the Kubelka–Munk ratio (10–12, 16):

$$\frac{K}{S} = F(R) = \frac{(1 - R)^2}{2R}$$

K/S may be simply viewed as an empirical transformation of reflectance that has a linear relationship with concentration. For this purpose, it would be simpler to use the nomenclature for the "Kubelka–Munk function" (17), F or $F(R)$, for only a single new variable is introduced. However, this nomenclature has not been used in the literature (10–12) on the

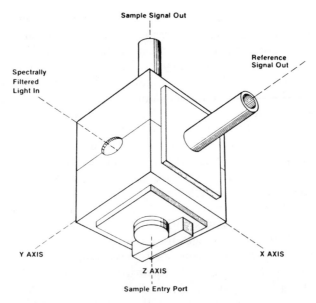

Figure 8.4. Configuration of the ports into the ''integrating sphere'' used to illuminate a test strip with optically diffuse light. Reprinted with permission from *Analytical Chemistry, 53*(12) 1950 (1981). Copyright 1981 American Chemical Society.

ARIS, and therefore K/S is used to denote this transformation throughout this text. For the interested reader, the derivation of K/S is given in Section 6.1.

Despite the fact that K/S varies linearly with the concentration of TMB, determination of concentration of drug is still quite complex. This is because the ARIS is a kinetic assay, rather than an end point assay. The presence of drug causes reactivation of apoenzyme, which in turn determines the *rate* of oxidation of TMB. Unfortunately, not even the rate of oxidation is constant because apoenzyme continues to be reactivated while TMB is being oxidized. Therefore, K/S versus time is not a straight line. Figure 8.5 shows curves for K/S versus time for theophylline.

Because of this complexity, timing of reflectance measurements must be precisely synchronized with the application of diluted specimen to the reagent strip. Concentration of drug is determined from the mean rate of change in K/S over a fixed time interval which begins at a critical time after sample is applied to the test strip. In the reports cited, the change in K/S between 60 and 80 s after sample was added was used in the assay for theophylline (11), and the change in K/S between 70 and 90 s was used to assay for phenytoin (12). For both drug assays, a plot of the change

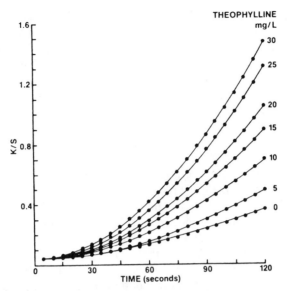

Figure 8.5. Kinetic response of the reagent strip of the apoenzyme reactivation immunoassay system (ARIS) for theophylline. The concentration of theophylline in the sample prior to dilution is shown for each curve. Reprinted with permission from *Clinical Chemistry, 31,* 738 (1985). Copyright 1981 The American Association for Clinical Chemistry.

in K/S, during the fixed time interval, versus drug concentration yields a straight line for drug concentrations in the range of 0 to 40 mg/L. Thus a two-point calibration can be used.

The Seralyzer® (Ames Division, Miles Laboratories, Inc., Elkhart, Indiana 45615) contains a microprocessor that automates reflectance measurements and performs all calculations necessary to arrive at a serum concentration. A different plug-in module for each assay contains the algorithm needed to perform that assay, as well as the interference filter for the light source (18). A sliding table facilitates introduction of the test strip into position for reflectance measurements after sample is applied. Timing of reflectance measurements begins when the operator presses a start button.

3.4. Test Procedure

The following test procedure assumes that the Seralyzer is the instrument being used. Into 800 μL of distilled water, 30 μl of patient serum is diluted. The operator pulls out the sliding reagent table and places a test strip in position on the table. Thirty microliters of the diluted serum is manually applied to the test strip, while the start button is pressed simultaneously. Immediately thereafter the reagent table is pushed into position for reflectance measurements. The instrument performs all subsequent functions, and the serum concentration is displayed digitally.

The instrument is calibrated using only two standards. The calibration is stable for at least 2 weeks (11, 12).

3.5. Applications

The ARIS offers several significant advantages over other assays: the convenience of having all dry reagents on a single test strip, the relative simplicity of the analyzer used, and the speed with which assays may be performed. Commercial proponents of the assay contend that this makes the assay ideal for "bedside testing." However, the test strips described for TDM have several drawbacks if used for this purpose. First, in assays thus far described for theophylline and phenytoin, serum or plasma is applied to the test strip, not whole blood. Thus centrifugation of the sample is required. Second, the serum or plasma must be diluted. Finally, errors may occur if the operator does not synchronize pipetting diluted serum onto the test strip with pressing the start button. Full automation of the procedure would solve the latter two problems. However, such an instrument would be considerably more complex, and therefore be more expensive and require more maintenance than the Seralyzer.

The manufacturer's evaluation of the Seralyzer phenytoin assay (12) yielded within-run and between-run coefficients of variation (CV) of 4% or less in the critical range of 10–20 mg/L. At 5 mg/L, the within-run CV was 5.6%, and the between-run CV was 6.3%. At 30 mg/L, both CVs were less than 3%. A comparison between the Seralyzer assay (y) and HPLC (x) yielded a regression line: $y = 1.03x + 0.33$, $s = 1.27$, $r = 0.99$, and $n = 92$. The manufacturer's evaluation of the Seralyzer theophylline assay (11) revealed a within-run CV of less than 6% and a between-run CV of 5% or less for concentrations ranging from 10 to 25 mg/L. Comparison with HPLC produced the regression line: $y = 1.08x - 0.4$, $r = 0.99$, $s = 1.11$, and $n = 110$.

Independent evaluations of the Seralyzer theophylline assay have also shown encouraging results. One group (19) reported within-run CVs of 3% or less for theophylline levels in the range 5–25 mg/L, a between-run CV of 5.9% at 15 mg/L, and a correlation coefficient of .95 when compared to HPLC. A CV between 4 and 5% at 15 mg/L was obtained by another group (20). Other authors report correlation coefficients of .98 (21) and .96 (22) when the Seralyzer theophylline is compared to HPLC, and a correlation coefficient of .99 compared to fluorescence polarization immunoassay (23). When Seralyzer theophylline assays were performed by emergency medicine residents after only "minutes" of training (24), a correlation coefficient of .91 was obtained, when results were compared to fluorescence polarization immunoassay. It should be noted that only a few samples in the toxic range are included in the correlation studies. The assertion that the Seralyzer tends to underestimate toxic theophylline levels has been made (25), although the data are not provided.

The ARIS would seem to be a logical choice for development of qualitative assays for drugs of abuse. The chief disadvantages of the ARIS in TDM, the need to dilute the sample and the need to time reflectance measurements accurately, are obviated in a qualitative assay. Serum or plasma could be applied to a test strip, and the latter could be visually inspected for the color change. A properly constructed set of test strips could yield a qualitative coma screen in a matter of minutes. Drugs detected by qualitative screening could then be immediately retested using appropriately diluted serum and a reflectance photometer for quantitative results.

4. ENZYME IMMUNOCHROMATOGRAPHY

The method of enzyme immunochromatography (26) requires no instrumentation and the sample can be serum, plasma, or whole blood. The assay does not depend critically on the activity of the enzyme, and thus the assay can be performed at ambient temperature, and stability of reagents is not a problem. Assays for whole blood are commercially available (Acculevel®, Syntex Medical Diagnostics, Syva Co., Palo Alto, CA 94304) for theophylline, phenytoin, phenobarbital, and carbamazepine.

4.1. Principle of the Assay

The assay consists of three components: (1) a long paper test strip, (2) enzyme reagent, and (3) developer solution. Antibody to the drug of interest is immobilized on the long paper test strip. The enzyme reagent contains drug conjugated to peroxidase mixed with glucose oxidase. If the assay is to be performed on whole blood, wheat-germ agglutinin or sheep antiserum to human erythrocytes is also added to the enzyme reagent, the purpose being to aggregate the red cells and thereby prevent them from migrating up the test strip. The developer solution contains the substrates for the enzymes, that is, glucose and 4-chloro-1-naphthol. The latter is the indicator.

Sample is first added to the enzyme reagent. The test strip is then placed upright, with the bottom of the strip immersed in the resulting solution. The liquid component of the solution (i.e., everything but cells and fibrin aggregates) migrates up the full length of the test strip by capillary action. As the liquid moves up the test strip, drug from the sample competes with drug–peroxidase conjugate from the enzyme reagent for binding sites on the immobilized antibody. If the sample contains no drug (Fig. 8.6A), then all of the drug–peroxidase conjugate will be bound to antibody near the bottom of the test strip. If the sample contains drug (Fig. 8.6B), it will occupy some of the binding sites at the bottom of the test strip, and some of the drug–peroxidase conjugate will be free to move higher up on the test strip. The height to which the drug–peroxidase conjugate migrates increases as the amount of drug in the sample increases. Glucose oxidase will distribute evenly over the entire length of the test strip.

The developer solution allows the operator to see the height to which the drug–peroxidase conjugate has migrated. When the test strip is immersed in the developer solution, glucose is oxidized, producing hydrogen peroxide, along the entire length of the test strip, because glucose oxidase is distributed homogeneously. However, oxidation of the indicator (4-chloro-1-naphthol) by hydrogen peroxide occurs only in the presence of the drug-peroxidase conjugate. Oxidation of the indicator produces a blue color. The height of the color produced is measured and translated to a concentration of drug in the sample.

4.2. Preparation of the Test Strip

One representative protocol for immobilization of antibody on the chromatography paper is as follows. Chromatography paper is activated by immersion in a solution of 1,1'-carbonyldiimidazole in dichloromethane, then washed thoroughly. After drying, the activated chromatography paper is immersed in a solution containing IgG antibody to drug, at pH 9.5, and incubated. Next, the chromatography paper is washed with standard buffer, pH 7.0, inactivated in a solution of polyvinyl alcohol, and dried. This protocol is by no means unique. Activation of the chromatography paper is accomplished equally well with any of the conventional agents, such as p-nitrophenylchloroformate or sodium metaperiodate.

If the test strips are cut with straight edges, the top of the area of color change will have an exaggerated meniscus shape, presumably due to faster capillary flow at the edge of the paper. If the test strips are cut with serrated edges, the meniscus shape can be converted to a rocket shape. The precision of the assay depends critically on reproducible measurement of the height of the color change, and this is better achieved when

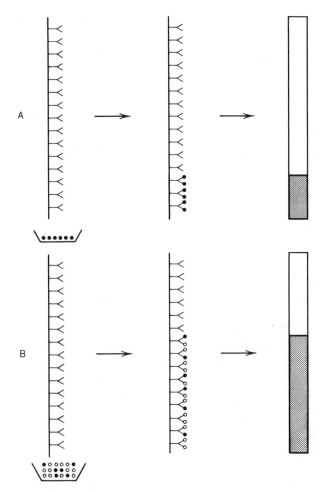

Figure 8.6. Schematic representation of enzyme immunochromatography. If no drug is present in the sample (A), all of the conjugate is bound at the bottom of the test strip. However, if drug is present in the sample (B), this drug will occupy binding sites at the bottom of the strip, and conjugate will be free to migrate higher. ○ = free drug; ● = drug conjugated to peroxidase.

the height of the apex of a rocket-shaped boarder is measured. Thus all test strips should be cut with serrated edges.

4.3. Test Procedure

Twelve microliters of sample is added to 1000 μl of enzyme reagent. The test strip is placed upright, with exactly 1 cm of the test strip immersed, in the sample/enzyme reagent solution. The solution is allowed to migrate up the length of the test strip, which takes about 10 min. Next, the entire test strip is immersed in developer solution for approximately 5 min. The height of the color change is measured, and the drug concentration is determined from a standard curve. The entire procedure is performed at ambient temperature.

The standard curve, once determined for a given lot of test strips, is purportedly stable. This is because quantification does not depend on enzyme activity, but rather on the affinity of the immobilized antibody. Therefore, for commercially produced products, the standard curve can be constructed when a given lot is manufactured, obviating the need for calibration by the user. When test strips for theophylline were stored desiccated at 4°C or 37°C, they remained stable during a 21-week trial, with a coefficient of variation of 5–8%.

4.4. Applications

The major advantages of immunochromatography are that no instrumentation and no calibration by the user are required. For this reason, the most logical application of the assay is for situations in which the test volume is too low to justify a capital expenditure for instrumentation. The physician's office is one setting in which this assay may be appropriate. However, the assay has several drawbacks: it involves the use of wet reagents, pipetting of sample and reagent, and several manual steps lasting approximately 20 min, although the actual hands-on time is minimal. Finally, there is an inherent lack of precision in the assay owing to the nature of the measurement; that is, a length in millimeters is converted to a serum concentration.

A multicenter evaluation (27) of the Acculevel immunochromatographic theophylline assay showed coefficients of variation (both within-run and between-run) ranging from 5% to 11% and correlation coefficients ranging from .93 to .97. These authors assert that the Acculevel theophylline assay is sufficiently "accurate" to replace laboratory methods. Because standard reference methods for theophylline have coefficients of variation of less than 5% (28–30), the precision of the Acculevel is less, but may be satisfactory for patient monitoring.

Immunochromatography is a potentially useful technique for emergency toxicological screening. A somewhat similar assay has been described for detection of morphine (31). A well-chosen panel of test strips could be developed for an emergency toxicology screen. It could be designed as a qualitative test, or the drugs detected by the screen could be quantitated. The test would be no more labor-intensive than, for example,

TLC methods now commonly employed for qualitative urine drug screens. The test would be simple enough that it could be performed quickly by any of the laboratory personnel, probably in less than 1 hr. This is in contrast to standard methods of serum toxicology screening now employed, such as HPLC or GC/MS.

5. FILM LAYER TECHNIQUES

The most technologically advanced of the dry reagent chemistries are those utilizing film layer techniques (18, 32–34). These techniques expand the number of assay systems that are amenable to a dry reagent format. The film layer technique allows chemical and physical processes to occur on separate layers, which essentially serve as separate compartments. Thus chemical reactions that would interfere with each other if performed in a single reaction chamber can still be incorporated on one slide. The film layer technique also allows reagents that must remain separate prior to use to reside on the same test slide. Development of these test slides was made possible by the ability to cast thin films precisely and by the ability to prepare quantitative chemical reagents within those films. Not surprisingly, much of this technology was developed in the photographic industry. The advent of color photography required that quantitative colorimetric chemistries could be performed in discrete film layers. An instant color print may consist of as many as 15 such layers.

5.1. Design of the Test Slide

The structure of the test slide (Fig. 8.7) consists of a transparent support layer on which one or more reagent layers are coated. Above the reagent layer(s) is a spreading layer. Sample is applied to the spreading layer and colorimetric changes in the reagent layer(s) are monitored through the transparent support layer.

The function of the spreading layer is to absorb the sample quickly and distribute it evenly over the surface of the reagent layer. The spreading layer typically consists of cellulose acetate pigmented with titanium oxide. The spreading layer is a highly porous structure, typically having a void volume of 80%. Fluid is rapidly and uniformly distributed throughout the void volume. Hydration of the gelatin that comprises the reagent layer occurs much more slowly. Therefore, entry of analyte into the reagent layer is limited by the intrinsic rate of hydration of gelatin, not by the delivery of fluid to the interface between the spreading layer and the reagent layer, and the amount of analyte that diffuses into the reagent

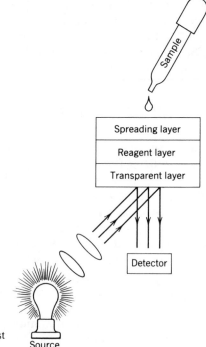

Figure 8.7. Architecture of a film layer test slide.

layer is relatively independent of the volume of fluid applied to the spreading layer. The net result is that the assay is relatively insensitive to the volume of sample applied to the slide. An error of 10% in the volume of the drop of sample applied to the slide results in less than a 1% error in the resulting reflection density.

The spreading layer serves two other purposes: it traps turbid or colored components in the serum that would interfere with measurements of reflectance density, and it acts as a white optical diffuser for incident light. In many assays, some of the reagents involved in the assay are incorporated in the spreading layer; in these assays, the spreading layer also serves as the first of the reagent layers.

Reagent layers consist of the chemical components of the assay incorporated in a gelatin binder. One or more of these layers are spread over polyethylene terephthalate film (i.e., the transparent plastic support layer). Other types of film layers that serve special functions (such as a semipermeable membrane) may be interposed between reagent layers.

5.2. Instrumentation and Data Handling

The colorimetric changes produced on the test slide are monitored by a reflectance densitometer. Light from a source bulb is passed through a system of filters and lenses to produce a narrow beam of light of a specific wavelength, which is oriented at an angle of 45° from the normal axis of the test slide. The spot of illumination produced on the slide is approximately 3 mm in diameter. The illumination is viewed along the normal axis. Reflectance data are expressed as the reflectance density,

$$D_r = \log \left(\frac{R_0}{R_t - R_f} \right)$$

where R_0 is the intensity of reflected light read by the detector with a standard reflector (a barium sulfate surface), R_t is the intensity measured with the test slide, and R_f is an empirical correction factor which accounts for flare and possibly other effects. Unfortunately, the relationship between concentration of product and reflectance density is not linear. Therefore, a transformed density, D_t, is defined by

$$D_t = -0.194 + 0.469 D_r + \frac{0.422}{1 + 1.179 \exp(3.379) D_r}$$

This transformation is based on that described by Williams and Clapper (35). The quantity D_t varies linearly with concentration. For the interested reader, this transformation is discussed in more detail in Section 6.2.

5.3. Examples of Toxicological Assays Using Film Layering Techniques

The Ektachem® theophylline assay (36) exploits the uncompetitive inhibition by theophylline of bovine hepatic alkaline phosphatase. The assay is performed at pH 8.2 so that endogenous alkaline phosphatase, which has a maximum activity at pH 10.5, will not interfere. The spreading layer contains the substrate, p-nitrophenyl phosphate. Sample diffuses from the spreading layer, carrying the substrate, into a layer of buffer, where the alkaline phosphatase in the patient's serum is inactivated. The sample and substrate then diffuse into the enzyme reagent layer, where the bovine hepatic alkaline phosphatase hydrolyzes the p-nitrophenyl phosphate. The rate of formation of product, p-nitrophenol, in the reagent layer is determined by measuring reflectance at a wavelength of 400 nm at two points in time, 1.17 and 2.5 min after addition of sample. The rate of formation of product is inversely proportional to the theophylline concentration.

In the assay under development for phenytoin (37), a limited number of antibody binding sites are immobilized in the spreading layer. A phenytoin analog conjugated to glucose oxidase is also present in the spreading layer. A phenytoin analog is used so that the antibody will not bind too avidly to the conjugate. Phenytoin in the patient's serum competes with the conjugate for binding sites in the spreading layer. Unbound conjugate diffuses into the reagent layer, where it produces a color change by a coupled enzyme assay with peroxidase.

5.4. Applications

The film layered test slides are used in fully automated instruments such as the Kodak Ektachem 400/700® analyzers, and a simpler instrument, the Ectachem DT60®. In the latter, the operator must manually place the appropriate test slide in the instrument and then apply sample to the slide. Manual steps cause less imprecision in film layer assays than in the other assays discussed because in the former, undiluted serum can be applied to the slide, and the result of the assay does not critically depend on the volume of the sample applied to the test slide. These unique advantages are afforded by the properties of the spreading layer, as described above.

Unfortunately, the only toxicological test offered commercially at the present time is theophylline. However, many other TDM assays are anticipated. On the other hand, because slides are available for numerous other nontoxicological clinical assays, purchase of the instrument need not be justified solely by the need for TDM.

6. MATHEMATICAL TRANSFORMATIONS USED IN REFLECTANCE SPECTROPHOTOMETRY

Derivations of the transformations used in reflectance spectrophotometry are given in this section.

6.1. The Kubelka–Munk Ratio

The collisions between light and particles of matter can be described in terms of two phenomena: absorption and scatter. The absorption coefficient, K, is the fraction of light absorbed (converted to heat) per unit length of medium. Note that this quantity is distinct from and should not be confused with the absorption coefficient A, used in Beer's law. The scattering coefficient, S, is the fraction of light scattered at an angle greater than 90° per unit length of medium.

Consider a homogeneous medium with a flat surface in the y–z plane, at $x = 0$. Let the positive direction of the x axis point into the medium. For simplicity, assume that the medium extends out to infinity in all directions in the y–z plane and in the positive x direction. The surface is illuminated by diffuse monochromatic light, that is, light of a given wavelength that radiates with equal intensity in all directions. Light that illuminates the surface must radiate at an angle from 0° to 90° with respect to the x axis. Conversely, light that is reflected from the surface must radiate at an angle between 90° and 180° with respect to the x axis. If all scattering is isotropic, that is, uniform in all directions, then at any point within the medium, light will radiate with equal intensity, I, in all directions at an angle less than 90° with respect to the x axis. Furthermore, light will radiate with equal intensity, J, in all directions at an angle greater than 90° with respect to the x axis. By symmetry, I and J must depend only on x.

Next, consider the change in the intensity of light through a slab of medium of thickness dx. The change in I is the sum of three terms: the loss of light due to absorption, the loss of light due to scattering, and the gain of light due to scattering. Therefore

$$dI = -KI\,dx - SI\,dx + SJ\,dx$$

A similar analysis is made for J, and therefore

$$-dJ = -KJ\,dx - SJ\,dx + SI\,dx$$

The negative sign in front of dJ is necessary because this radiation has a net flux in the minus x direction. Solution of these equations yields:

$$I = I_0 \exp(-wx) + I_1 \exp(wx)$$

$$J = J_0 \exp(-wx) + J_1 \exp(wx)$$

where $w = (K^2 + 2KS)^{1/2}$

$$J_0 = \left(-\frac{w}{S} + \frac{K}{S} + 1\right) I_0$$

$$J_1 = \left(\frac{w}{S} + \frac{K}{S} + 1\right) I_1$$

Because I and J must approach 0 as x approaches infinity, $I_1 = J_1 = 0$. At the surface of the medium ($x = 0$), $I = I_0$ and $J = J_0$. By definition, the reflectance is given by J_0/I_0, so that

$$R = \frac{J_0}{I_0} = -\frac{w}{S} + \frac{K}{S} + 1$$

Substituting for w and simplifying yields

$$\frac{K}{S} = \frac{(1 - R)^2}{2R}$$

It is quite remarkable that the reflectance depends only on the ratio, K/S, and not the individual parameters. It should be noted that K and S vary independently with the wavelength of incident light. Therefore the ratio K/S varies for different wavelengths of light.

It is borne out by experimental observation (38), and is perhaps intuitive, that for a low concentration of a pigment embedded within a white matrix, the absorptivity, K, varies linearly with the concentration of the pigment, while the scattering, S, remains constant. Therefore K/S varies linearly with the concentration of pigment. Note that K/S increases as the reflectance, R, decreases. Thus if a compound causes absorption of light of a given wavelength, presence of that compound causes K/S to increase at that wavelength, and less light is reflected.

6.2. The Williams–Clapper Transformation

Consider a beam of light incident at an angle of 45° with respect to the normal upon a diffuse reflector coated by a thin, uniform layer of gelatin. Let the intensity of light reflected along the normal axis be R_t. Next, consider an identical beam, incident at 45°, on a perfect white diffuse reflector. Let the intensity of light reflected along the normal axis from this surface be R_0.

First calculate the attenuation of light as it traverses the path shown in Fig. 8.8. This is

$$A = \frac{(0.945)(0.956)}{(1.53)^2} R_b T^{2.13}$$

The factor 0.945 accounts for a 5.5% loss in intensity when light enters the gelatin from air, at 45°. The factor 0.956 accounts for a 4.4% loss in

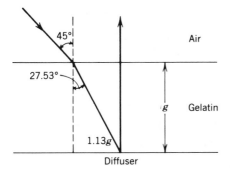

Figure 8.8. Path of light reflected from an optical diffuser coated by a layer of gelatin. Internal reflection at the gelatin–air surface is ignored. (g equals the thickness of the gelatin.)

intensity for light exiting the gelatin into air along the normal axis. These factors are calculated from the well-known Fresnel equations (39). R_b is the reflectance of the base beneath the gelatin. T is the transmittance for light passing through the entire thickness of the gelatin along an axis normal to its surface. The exponent 2.13 arises as follows: light incident on the gelatin–air surface at 45° is refracted to 27.53° by the law of sines, with the index of refraction being 1.53 for gelatin and 1.00 for air. The path length for incident light to reach the base is therefore the secant of 27.53°, or 1.13, times the thickness of the gelatin. The total path length for light entering and exiting the gelatin is therefore 2.13 times the thickness of the gelatin.

Finally, the intensity of light along the normal axis is attenuated by the square of 1.53 owing to refraction of light as it exits the gelatin; this is shown by the following analysis. Consider the flux of light reflected from a perfect white optical diffuser within an angle α of the normal axis. If this diffuser is now coated with a thin layer of gelatin, and no reflection at the gelatin–air interface or absorption occurs, then the same amount of flux will be spread over a wider angle, β, with respect to the normal axis. At a distance h that is very large compared to the thickness of the gelatin, the unrefracted light is spread over a circle of radius $h \tan \alpha$, and similarly, the refracted light is spread over a circle of radius $h \tan \beta$. Because the flux in each case is the same,

$$\pi(h \tan \beta)^2 I_r = \pi(h \tan \alpha)^2 I_u$$

where I_r and I_u are the intensities of refracted and unrefracted light, respectively. Because for small angles the tangent is equal to the sine, and using the law of sines,

$$\frac{I_r}{I_u} = \frac{\tan^2 \alpha}{\tan^2 \beta} = \frac{\sin^2 \alpha}{\sin^2 \beta} = \frac{1}{n^2}$$

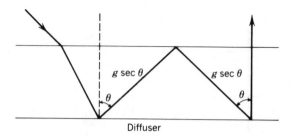

Figure 8.9. Path of light internally reflected once at the gelatin–air surface (g = gelatin thickness).

Some of the light reflected from the diffuser at an angle θ with respect to the normal axis is internally reflected from the gelatin–air interface; this light reilluminates the diffuser at the base, and therefore adds to the light reflected along the normal axis (Fig. 8.9). The fraction of light internally reflected, $R(\theta)$, is defined by the Fresnel equations (39). The light is attenuated by the gelatin. The base-to-base path length for internally reflected light is 2 sec θ times the thickness of the gelatin, so this quantity is the exponent of T. R_b accounts for the fact that internally reflected light is again attenuated by the diffuser. The fraction of light that reilluminates the base is the component of light parallel to the normal axis, averaged over the entire hemisphere. Using the mean value theorem, this fraction is given by

$$X = \frac{\int R(\theta)T^{2\sec\theta}R_b \cos\theta \sin\theta \, d\theta \, d\phi}{\int \cos\theta \sin\theta \, d\theta \, d\phi}$$

$$= 2R_b \int_0^{\pi/2} R(\theta)T^{2\sec\theta} \cos\theta \sin\theta \, d\theta$$

Because light may be internally reflected an infinite number of times (Fig. 8.10), R_t/R_0 is the sum of a geometric series:

$$\frac{R_t}{R_0} = A + AX + AX^2 + \cdots = \frac{A}{1 - X}$$

or

$$\frac{R_t}{R_0} = \frac{(0.945)(0.956)}{(1.53)^2} R_b T^{2.13} \left[1 - 2R_b \int_0^{\pi/2} R(\theta)T^{2\sec\theta} \cos\theta \sin\theta \, d\theta \right]^{-1}$$

This is the Williams–Clapper transformation.

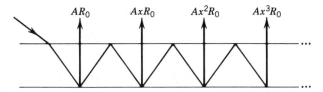

Figure 8.10. Contribution of multiple internal reflections to the intensity of light reflected from the gelatin-coated diffuser.

The integral must of course be solved numerically. By writing $D_r = -\log (R_t/R_0)$ and $D_t = -\log (T)$, and solving for D_t, the transformation equation given in Section 5.2 is derived. Note that D_t varies linearly with concentration because the transmittance of the gelatin itself does follow Beer's law. Also note that in this theoretical analysis, it is assumed that the correction factor $R_f = 0$.

REFERENCES

1. M. E. Geffner, S. A. Kaplan, B. M. Lippe, and M. L. Scott, *J. Am. Med. Assoc., 249,* 2913 (1983).
2. R. B. Tattersall, *Diabetologia, 16,* 71 (1979).
3. B. Barr, S. B. Leichter, and L. Taylor, *Diabetes Care, 7,* 261 (1984).
4. G. T. Gillett, *Lancet, 2,* 209 (1986).
5. T. M. Li, J. L. Benovic, R. T. Buckler, and J. F. Burd, *Clin. Chem., 27,* 22 (1981).
6. A. C. Greenquist, B. Walter, and T. M. Li, *Clin. Chem., 27,* 1614 (1981).
7. B. Walter, A. C. Greenquist, and W. E. Howard, *Anal. Chem., 55,* 873 (1983).
8. W. E. Howard, A. C. Greenquist, B. Walter, and F. Wogoman, *Anal. Chem., 55,* 878 (1983).
9. D. L. Morris, P. B. Ellis, R. J. Carrico et al., *Anal. Chem., 53,* 658 (1981).
10. R. J. Tyhach, P. A. Rupchock, J. H. Pendergrass et al., *Clin. Chem., 27,* 1499 (1981).
11. P. Rupchock, R. Sommer, A. Greenquist, R. Tyhach, B. Walter, and A. Zipp, *Clin. Chem., 31,* 737 (1985).
12. R. Sommer, C. Nelson, and A. Greenquist, *Clin. Chem., 32,* 1770 (1986).
13. H. R. Schroeder, P. K. Johnson, C. L. Dean, D. L. Morris, D. Smith, and S. Refertoff, *Clin. Chem., 32,* 826 (1986).
14. D. L. Morris, *Anal. Biochem., 151,* 235 (1985).
15. M. A. Genshaw and R. W. Rogers, *Anal. Chem., 53,* 1949 (1981).

228   DRY REAGENT CHEMISTRIES

16. G. Kortum, *Reflectance Spectroscopy,* Springer, New York, 1969, pp. 219 ff.
17. *Ibid.,* pp. 106–116.
18. B. Walter, *Anal. Chem., 55,* 499A (1983).
19. R. Lindberg, K. Ivaska, K. Irjala, and T. Vanto, *Clin. Chem., 31,* 613 (1985).
20. F. M. Cuss, J. B. Palmer, P. J. Barnes, *Brit. Med. J., 291,* 384 (1985).
21. A. Morikawa, K. Tajima, M. Mitsuhashi, K. Tokuyama, H. Mochizuki, and T. Kuruome, *Ann. Allergy, 55,* 72 (1985).
22. G. Z. Lotner, G. E. Vanderpool, M. S. Carroll, D. L. Spangler, and D. G. Tinkelman, *Ann. Allergy, 55,* 454 (1985).
23. R. P. Busch and M. A. Virji, *Clin. Chem., 31,* 1247 (1985).
24. T. Reinecke, D. Seger, and R. Wears, *Ann. Emerg. Med., 15,* 147 (1986).
25. L. Amdall, G. G. Shapiro, C. W. Bierman, C. T. Furukawa, and W. E. Pierson, *J. Allergy Clin. Immunol., 75* Suppl., 127 (1985), abstract 91.
26. R. F. Zuk, V. K. Ginsberg, T. Houts et al., *Clin. Chem., 31,* 1144 (1985).
27. L. M. Vaughan, M. M. Weinberger, G. Milavetz et al., *Lancet, 1,* 184 (1986).
28. J. Chang, S. Gotcher, and J. B. Gushaw, *Clin. Chem., 28,* 361 (1982).
29. H. D. Hill, M. E. Jolly, C. H. J. Wang et al., *Clin. Chem., 27,* 1086 (1981), abstract.
30. G. W. Peng, M. A. F. Gadalla, and W. L. Chiou, *Clin. Chem., 24,* 357 (1978).
31. D. J. Litman, R. H. Lee, H. J. Jeong et al., *Clin. Chem., 29,* 1598 (1983).
32. H. G. Curme, R. L. Columbus, G. M. Dappen et al., *Clin. Chem., 24,* 1335 (1978).
33. R. W. Spayd, B. Bruschi, B. A. Burdick et al., *Clin. Chem., 24,* 1343 (1978).
34. T. L. Shirey, *Clin. Biochem., 16,* 147 (1983).
35. F. C. Williams and F. R. Clapper, *J. Opt. Soc. Am., 43,* 595 (1953).
36. Theophylline Test Methodology, Eastman Kodak Co., Rochester, NY, 1986.
37. S. J. Danielson, R. Olyslager, P. C. Hosimer, and M. W. Sundberg, unpublished data presented at the American Association of Clinical Chemistry meeting, Chicago, 1986.
38. G. Kortum, *Reflectance Spectroscopy,* Springer, New York, 1969, p. 180.
39. *Ibid.,* pp. 5–21.

CHAPTER

9

DRUG OF ABUSE PROFILE: COCAINE*

PETER I. JATLOW

Yale University School of Medicine
New Haven, Connecticut

1. HISTORY

Cocaine is of plant origin (1–4), as are many psychotropic substances with a long history of human consumption. It is the major alkaloid of the plant species *Erythroxylon coca*, which is widely cultivated throughout South America in the foothills of the Andes and in the Amazon regions. Chewing coca leaves to self-administer cocaine was important in the Incan and pre-Incan Andean civilizations more than a thousand years ago, and there is archaeological evidence of coca leaf chewing at least as early as 3000 B.C.

Cocaine as coca was introduced to Europe in the sixteenth century. The active principle was isolated from the coca leaf by Albert Niemann

* This chapter under the same title, was originally published by the American Association of Clinical Chemistry in the Proceedings of the Arnold O. Beckman Conference in Clinical Chemistry, *Clin. Chem.*, *33* No.(11B), 66B–71B (1987) and, with minor modifications, is reprinted here, with their permission.

around 1860. Around 1923, Richard Willstatter determined the structure of cocaine and accomplished its synthesis. Subsequently Carl Koller reported its value as a local anesthetic, a purpose for which it is indeed still used in conjunction with the intubation of surgical patients. Freud, initially a strong proponent of cocaine, was clearly intrigued by this drug, advocating its use as a stimulant for the treatment of gastric disorders, morphine and alcohol addiction, and asthma, and as an aphrodisiac.

Freud may have been prescient when he predicted a "very great future" for cocaine use in America, although it seems doubtful that he anticipated the negative consequences (1). In any event, at the turn of the century cocaine enjoyed a brief surge of popularity as a component of stimulants, tonics, and sodas, including Coca Cola (this active ingredient has been excluded from the Coca Cola formulation since 1903). The potential for abuse of cocaine was recognized fairly early and, despite some controversy, its sale was banned by the Harrison Act in 1914.

The era of cocaine use in the United States, barely 100 years, has been a relatively brief period in the long history of this drug. However, an increase in cocaine use, changes in patterns of use and dosage in association with increased morbidity and mortality, and a heightened appreciation of the consequences of cocaine abuse have mandated an improvement in our understanding of cocaine's effects and mechanisms of action. A 1982 survey indicated that approximately 4,000,000 Americans had used cocaine within the last 30 days. There has been a moderate increase in the prevalence of use since then. More dramatic has been the marked increase in the reported clinical consequences of cocaine use, as reflected by fatal overdoses, emergency room visits, and applications to treatment programs (5, 6).

2. CLINICAL PROPERTIES AND TOXICITY

Far more is known about what cocaine does than about the mechanisms of its actions. Cocaine is a local anesthetic, has sympathomimetic properties, and is a strong stimulant of the central nervous system (CNS). Most of its known pharmacological actions are related to these properties (2, 7–9). Cocaine's local anesthetic and vasoconstrictive properties are the basis for its accepted medical use prior to nasal intubation, and as an adjunct during otorhinolaryngeal surgery. Indeed, the plasma concentrations of cocaine achieved after these clinical applications are similar to those associated with significant psychotropic effects in different settings (10). They appear not to be associated with significant CNS stimulation or pleasurable effects in the operating room environment, when administered in conjunction with premedication. In common with other sympathomimetic agents, cocaine induces a significant increase in blood pressure, pulse, and less dramatically, body temperature. Acute CNS effects after a single dose include an increased sense of alertness and the pleasurable sensation defined as euphoria.

The psychotropic effects of cocaine occur more rapidly and intensely after its instantaneous administration by intravenous or smoking routes. Cocaine administration, especially heavy use, is often followed by a "crash," associated with irritability, depression, dysphoria, and physical discomfort. Cocaine is one of the most reinforcing of drugs, in that animals given free access to cocaine will self-administer it to the point of debilitation or death (11, 12). The compulsive smoking of cocaine free base ("crack") may represent a similar phenomenon.

In common with other sympathomimetic agents, cocaine is reported to produce acute tolerance to its affects on the cardiovascular system and behavior (13–17). At the same plasma concentrations, the behavioral effects are greater when concentrations are increasing than when they are decreasing, producing a characteristic clockwise hysteresis curve when concentration is plotted versus effect over time (14, 15). After an initial dose of cocaine, subsequent doses produce a significantly lesser effect (13). Chou, Ambre, and Ruo (17) have analyzed cocaine's pharmacodynamics kinetically and quantified the rate of development of tolerance to the chronotropic effects of cocaine.

The acute toxicity of cocaine can be seen as an extension of the effects described above. Acute overdoses, including those associated with fatalities, may be associated with cardiac arrhythmias and symptoms of CNS stimulation progressing to seizures. Psychoses (paranoia, hallucinations) may also be seen in acute overdose as well as in chronic toxicity. Delirium with hyperpyrexia has also been reported after cocaine overdose, although to what extent this complication is a direct effect of cocaine or a consequence of seizures is not clear (18). An increasing number of cases of acute myocardial infarction associated with cocaine use has been reported (19, 20). In some of these cases, the coronary arteries have been described as normal. In these instances, the infarct has been thought to be a consequence of acute coronary artery spasm, or an acute arrhythmia. However, the possibility that chronic cocaine use may be associated with accelerated coronary arteriosclerosis and occlusion has been raised in a recent report of coronary vascular disease in relatively young cocaine users who had experienced myocardial infarcts (20). The question of cause and effect has not been resolved: we do not know whether individuals with preexisting coronary disease are more susceptible to the consequences of cocaine use, such as an acute increase in blood pressure and pulse.

Cocaine causes acute hepatic necrosis in sensitive strains of mice (21–27), but evidence that it can cause liver damage in humans is as yet lacking. Hepatic toxicity in mice has been attributed to the formation of chemically reactive microsomal metabolites derived from norcocaine (24, 27). There appears to be species and strain selectivity regarding the sensitivity to cocaine's hepatotoxicity in rodents, but hepatic necrosis can be induced by cocaine in some ordinarily nonsensitive animals by pretreatment with a microsomal inducer such as phenobarbital (21). The putative toxic metabolite(s) are produced via oxidative microsomal pathways, which appear to be relatively minor for cocaine in humans, but more significant in rodents. However, with current new trends involving abuse of much larger doses of cocaine, the possibility that such alternative pathways are important (analogous to acetaminophen toxicity) should be considered. Because many cocaine abusers are polydrug abusers and at high risk for viral hepatitis, establishing causality in humans could be difficult. The hepatotoxicity of cocaine has been reviewed (24).

The recent dramatic increase in cocaine smoking, fueled by the easy availability of the free base in the form of "crack," raises the possibility of pulmonary toxicity. Several reports have described pulmonary diffusion defects in cocaine smokers (28, 29). The assumption is that cocaine's vasoconstrictive effects on the pulmonary vasculature play a role in its pulmonary toxicity, although direct toxic effects of cocaine on the lungs have not been excluded. Toxic effects to other organ systems as a consequence of its vasoconstrictive and hypertensive properties remain a possibility.

3. ROUTE OF ADMINISTRATION

3.1 Oral Administration

Despite its popularity as a constituent of wines and sodas at the turn of the century, the stability and bioavailability of cocaine administered orally were questioned until fairly recently. Data from our laboratory and elsewhere have established that cocaine administered in capsule form results in plasma concentrations and area under concentration versus time curves comparable with those obtained after intranasal administration, with equivalent psychological effects (30, 31). After oral administration and a lag phase of about 30 min, absorption is rapid, reaching peak plasma concentrations in about 60 min. Nonetheless, the oral route is not a very prevalent mode of administration among current drug abusers.

3.2 Buccal and Nasal Route

Historically, chewing coca leaves was the major route of administration of cocaine and has remained so among much of the native Andean population. Although cocaine was thought to reach the systemic circulation when coca leaves were chewed, not until recently have studies confirmed that this mode of administration results in concentrations of cocaine in plasma (32, 33) that are consistent with pharmacological activity. Probably both buccal (mucosal) and lower gastrointestinal tract absorption occur when coca leaves are chewed. Since antiquity, native chewers have added an alkaline material known as *tocra* or *llipta* during mastication of the leaves, apparently because the desired effect was enhanced by this maneuver. We have determined that the same quantity of coca leaves chewed with alkali resulted in substantially higher concentrations of cocaine blood levels than when the leaves were chewed alone. Conversion of cocaine to the un-ionized free base at an alkaline pH would of course facilitate its absorption across mucosal membranes. Cocaine has a pK_a of 8.6, and llipta or tocra produces a pH of 10–11. Thus an intuitive or empirical judgment by early and current coca leaf chewers appears to have had a sound scientific rationale.

Nasal insufflation (snorting) has remained the most popular method for recreational administration of cocaine, although smoking is increasingly popular. Absorption of cocaine by this route begins almost instantaneously but peak concentrations are not reached for 30–60 min, somewhat later than the onset of observed psychological effects. Mucosal absorption of cocaine is variable and complex, probably as a consequence of local vasoconstriction. Systemic bioavailability by this route has been estimated to be anywhere from 20% to 60% based on comparisons of area under the time versus concentration curves for the nasal versus the intravenous route, but obtained in different experiments with different sets of subjects (31, 34).

3.3 Intravenous and Pulmonary Route

Administration of cocaine by the intravenous (34–37) and smoking (38, 39) routes is comparable in that both routes produce almost instantaneous access to cocaine by the systemic circulation and brain. The large surface of area of the lungs permits rapid absorption through a noninvasive route. Cocaine paste, a crude form of cocaine of variable composition, has been smoked, mixed with tobacco, by urban South American youths for years (40). We have measured cocaine concentrations in plasma from paste smokers, and indeed, high concentrations are achieved in minutes (38). Cocaine free base, on the other hand, can be volatilized at lower temperatures, and with less decomposition than can the salt form. Thus smoking free base is more efficient than smoking paste for delivering cocaine to the systemic circulation. Preparation of free base had previously involved extraction of cocaine from aqueous solution into an organic solvent such as ether. More recently, direct precipitation of the free base ("crack") after alkalinizing an aqueous solution of the salt has resulted in easier and less hazardous preparation of the free base and resulted in its wider availability. Smoking of cocaine appears to be particularly reinforcing behavior associated with compulsive use, binges, and more intense withdrawal symptoms, and appears to accelerate the addictive process (41, 42).

4. METABOLIC DISPOSITION AND KINETICS

A knowledge of the kinetics and metabolism of cocaine is critical to understanding its clinical pharmacology and to the appropriate design of research protocols. Above and beyond these considerations, information regarding cocaine disposition must be taken into account in the development and selection of analytical methodology for measuring and identifying of cocaine in biological fluids and interpreting the data so obtained.

Most studies of cocaine kinetics have been based on relatively small numbers of subjects, but the data appear to be in reasonable agreement regardless of the route of administration (Table 9.1) (16, 17, 31, 34, 43–59). With an elimination half-life of approximately 1 hr, failure to detect the parent drug in blood or urine does not exclude fairly recent use. After

Table 9.1. Pharmacokinetic Parameters of Cocaine and Its
Major Metabolites[a]

	$T_{1/2}$ (hr)	V_d (L/kg)	Cl (L/min)
Cocaine	0.5–1.5	2	2
Benzoylecgonine	5–8		
Ecgonine methyl ester	3.5–6		

[a] $T_{1/2}$, V_d, and Cl are half-life, volume of distribution, and clearance, respectively.

oral and intranasal administration, up to a maximum dose of 2 mg/kg of body weight, the kinetics of cocaine elimination appear to be dose independent (31). The K_m values of approximately 50 and 1000 μmol/L reported by Stewart et al. (54) for plasma and liver esterases, respectively, suggest that saturation of these enzymes would be unusual. On the other hand, Barnett, Hawks, and Resnick (34) have reported evidence for dose-dependent kinetics after the intravenous administration to four subjects of 1–3 mg of cocaine per kilogram of body weight.

Cocaine is rapidly hydrolyzed by plasma and liver esterases to ecgonine methyl ester (EME), one of cocaine's two major metabolites (43, 44–46, 53, 54). EME accounts for about 30–50% of cocaine's urinary disposition, and benzoylecgonine for most of the remainder. Relatively little (about 1–5%) unchanged cocaine is excreted in the urine. Serum cholinesterase (pseudocholinesterase; EC 3.1.1.8) appears to be responsible for the hydrolysis of cocaine in blood (49, 53, 54). The rapid in vitro disappearance of cocaine in serum does not occur in samples from subjects who are homozygous for the atypical esterase enzyme as identified by low dibucaine numbers (49, 54). In vitro hydrolysis of cocaine in plasma is also inhibited by physostigmine and fluoride ion. It has not been established whether individuals with the atypical esterase show significantly impaired disposition of cocaine or are more sensitive to cocaine toxicity. However, it would seem prudent to determine cholinesterase activity and to phenotype research subjects before administering cocaine to them. Stewart et al. (54) have reported that hepatic enzyme systems are also capable of hydrolyzing cocaine to ecgonine methyl ester. They have measured the kinetic parameters for the plasma and liver esterases, and report that the liver enzyme has a lower affinity but higher capacity for cocaine substrate than the plasma enzyme has.

Evidence for enzymatic hydrolysis of cocaine to benzoylecgonine has not been demonstrated. Cocaine, however, does slowly hydrolyze to benzoylecgonine in aqueous solution at pH 7.4. Stewart et al. (54) found that cocaine incubated at pH 7.4 in buffered saline was 42% converted to benzoylecgonine; they suggest this is sufficient to explain the urinary excretion of this metabolite. One study reported in vitro production of benzoylecgonine by liver microsomes, but chemical hydrolysis was not ruled out (55).

The biological half-lives of EME and benzoylecgonine, approximately 4 and 6 hr, respectively (45, 46), account for their prolonged presence in urine, as compared with that of cocaine. Immunological methods, designed for screening for drugs of abuse, are directed at detecting benzoylecgonine. However, the ability to identify both metabolites in urine provides useful supporting information. Ambre (46), who emphasized the importance of EME as a urinary metabolite, suggests that the relative proportions of cocaine, EME, and benzoylecgonine in the urine may be helpful in establishing the time elapsed since administration. In any event, EME and benzoylecgonine account for more than 80% of the urinary disposition of cocaine, and are the metabolites most relevant to drug abuse screening programs. As measured with the usual methodology, they are detectable for 24–60 hr after cocaine use as compared with about 3–6 hr for the parent compound.

N-Demethylation of cocaine to norcocaine has been identified as a minor metabolic pathway in humans (48). However, to date norcocaine is the only cocaine metabolite found to have in vivo pharmacological activity in animals and to block amine reuptake in vitro (47). Norcocaine appears to be unimportant as a determinant of cocaine's behavioral effects in humans. However, metabolites of norcocaine such as N-hydroxynor-cocaine and norcocaine nitroxide have been suggested as mediators of cocaine's hepatotoxicity in animals (24).

Arylhydroxy and hydroxymethoxy metabolites of cocaine have also been identified in urines of cocaine users (51, 52). Interestingly, ethyl ester homologs of these compounds, presumably arising by transesteri-fication, have been detected in the urine of individuals using cocaine and ethanol concurrently (56).

Most of the data on the kinetics and metabolic disposition of cocaine so far have been derived from individuals who have been administered relatively modest single doses of cocaine in the absence of other drugs. Alternative pathways may become quantitatively more significant in me-gadosage administration, as is prevalent during free base binging. The effect of microsomal enzyme inducers such as phenobarbital or ethanol on the metabolic disposition of cocaine has not been evaluated in humans. It is also conceivable, but not established, that microsomal pathways might account for a greater proportion of cocaine's metabolic disposition in individuals with low cholinesterase activity or who have inherited the atypical enzyme.

5. DOSE, PLASMA CONCENTRATION, AND EFFECTS

The acute tolerance that develops during cocaine use makes it difficult to define a predictable relationship between plasma levels and response (13, 14, 16, 17). Rate of increase may be more important than absolute peak concentrations in plasma and rapidly increasing plasma concentra-tions appear to be associated with a greater response (16). Consequently, effects appear to be a function of route and rate of administration as well as total dose, the most intense effects being associated with intravenous administration or free base smoking. Mayersohn and Perrier's evaluation of data from our laboratory indicated a linear relationship between response (high) and log plasma concentration during the ascending portion of the plasma concentration curve, following intranasal administration of cocaine (50). Holford and Sheiner (15), analyzing the same data, demonstrated a clockwise hysteresis curve typical of acute tolerance. Oral and intranasal doses of about 0.4 mg/kg (30–40 mg) were associated with a peak plasma concentration of about 50 ng/mL, and with detectable euphoria (14, 30)—a value probably consistent with a single intranasal street dose. Intranasal doses of 1–2 mg/kg are associated with peak concentrations of 100–200 ng/mL. Cardiovascular effects and euphoria, as might be expected, generally peaked sooner than drug concentrations (14, 16). After intravenous injection of cocaine, dose-related changes in mood, pulse, and blood pressure occur in minutes. Fishman et al. (36) noted that significant changes were produced by administration of 8 mg intravenously, and that the effect of 16 mg appeared to be consistent with the usual street doses. Subsequently, they reported the development of acute tolerance for the psychotropic and cardiovascular effects (13). In a kinetic study of cocaine's effects, Chou, Ambre, and Ruo (17) reported concentrations in plasma of approximately 200 ng/mL, with pulse rate changes of about 60 beats/min after intravenous administration of 32 mg. Paley et al. (38), in a study of Peruvian paste smokers, reported peak concentrations in the range of 220–700 ng/mL after the smoking of cocaine paste, 3–5 mg/kg of body weight. Most likely, the use of comparable amounts of pure free base would have produced much higher concentrations. Plasma concentrations are still undoubtedly fairly high during the crash or dysphoric phase that can occur when the subject stops smoking.

A wide range of cocaine concentrations have been measured in fatal overdoses (57, 58). Wetli and Wright (57) reported concentrations ranging from 100 to 17000 ng/mL (the lower concentrations may have been artifactually altered by postmortem hydrolysis). It has not been established whether the cardiac complications, including arrhythmias and acute myocardial infarction, associated with cocaine use are dose- or plasma-concentration-dependent.

6. NEUROCHEMICAL MECHANISMS

In common with other local anesthetics, cocaine blocks sodium transport across membranes. However, there is general agreement that the physiological and psychotropic effects commonly associated with cocaine use are probably not a consequence of its local anesthetic properties. It also appears unlikely that cocaine's effects can be entirely attributed to actions on any one or two neurochemical systems. Indeed, cocaine has been reported to interact with norepinephrine, dopamine, serotonin, acetyl-choline, and γ-aminobutyric acid neuronal systems (9, 59).

Cocaine in vitro blocks reuptake of norepinephrine into presynaptic nerve terminals. Its characteristic effects on blood pressure, heart rate, pupil size, and body temperature are consistent with augmented norep-inephrine activity. Cocaine also blocks reuptake of dopamine and sero-tonin. Increased concentrations of monomines at the synaptic cleft with binding to presynaptic inhibitory autoreceptors may also affect activity of monomine neurons. In animals, dopamine systems seem to play an important role in mediating the reinforcing actions of cocaine and opiates (60, 61). Lesions of dopamine neurons and large doses of dopamine block-ing drugs reduce drug seeking behavior. In humans the mechanisms of drug-induced euphoria and pharmacological reward are also thought to involve dopamine systems. On the other hand, animals undergoing chronic cocaine administration show evidence of catecholamine deple-tion. The roles of amine depletion and/or changes in receptor sensitivity in the development of acute tolerance to cocaine and withdrawal symp-toms remains to be determined.

The search for cocaine receptors has led to the identification of sat-urable binding sites on serotonergic and dopaminergic neurons (62, 63) with affinity constants reported to be in the low micromolar range. Binding appears to be associated with amine transport sites in some but not all instances.

7. ASSAY OF COCAINE AND METABOLITES

The analytical technology for identification or quantitation of cocaine and its major metabolites is reasonably well advanced. As with other drugs of abuse, selection of methodology should take into account the purpose for which the assay is performed, being constrained by the specifics of cocaine's pharmacokinetics and metabolic disposition.

For purposes of correlating blood (plasma or serum) concentrations of cocaine with clinical or behavioral effects, analytical methods should be capable of measuring plasma concentrations as low as 50 ng/mL. Phar-macokinetic studies should involve methods that are capable of measuring concentrations less than 10 ng/mL. Most studies to date have been based on gas chromatography with nitrogen selective detection, a technique that is capable of measuring concentrations in the 2–5 ng/mL range (64, 65).

Derivatives that elicit a response from electron capture detectors have also been prepared (66). Liquid chromatographic methods for measurement of cocaine have also been reported but are to date still somewhat less sensitive than gas chromatography (68, 69). The lower limit for quantitation of cocaine and its metabolites in biological fluids appears to be about 20–50 ng/mL for liquid chromatography. However, the greater ease of use of liquid chromatography and its potential application to the more polar metabolites of cocaine make it a tempting alternative despite its slightly lesser sensitivity. Gas chromatography/mass spectrometry (GC/MS) procedures, both electron impact and chemical ionization, have been developed for the analysis of cocaine and its metabolites in plasma (70–72). As with other drugs, this technique remains the state of the art for quantitation, especially when deuterated internal standards are used.

Norcocaine, a minor but apparently active metabolite of cocaine has also been measured with GC/MS (47, 72).

Benzoylecgonine, a major urinary metabolite of cocaine, is also found in plasma. Its measurement presents a challenge because of the compound's water solubility. Successful extraction of benzoylecgonine requires relatively nonselective solvents systems such as chloroform/alcohol mixtures. Although these provide good recovery of benzoylecgonine, the resulting chromatograms may be relatively "dirty" or complex, especially if additional clean-up steps are not incorporated into the procedure. Nevertheless, good GC and GC/MS procedures have been described for detecting and quantifying benzoylecgonine (70, 71). Generally, derivatization of benzoylecgonine is required for adequate GC analysis. The various derivatization techniques described usually involve esterification of the carboxyl group (67, 70, 71). In some procedures benzoylecgonine has been re-esterified to cocaine, although preparation of other derivatives would appear to be preferable. Reverse phase "high pressure" liquid chromatography is an ideal alternative for measuring benzoylecgonine because this technique does not require derivatization (68). Benzoylecgonine has reasonably good absorbance at 235 nm. For purposes of screening for drug abuse, excellent radioimmunoassays and nonisotopic immunoassays are available for detecting benzoylecgonine in the urine (73, 74). As with other immunoassays, positive findings should be confirmed with a more definitive procedure, ideally GC/MS.

Recently, interest has focused on the measurement of EME produced by the action of plasma and hepatic esterases on cocaine. Like benzoylecgonine, significant concentrations of EME are found in the urine and it has been measured by GC/MS (44–46). Although it can be quantified by GC without derivatization, we have found that acetylation improves chromatographic behavior, at least when packed columns are employed.

Loss of the aromatic ring from cocaine appears to preclude the sensitive detection of EME, at least in its underivatized form, by liquid chromatography.

Thin-layer chromatography (TLC) should not be neglected as a useful technique for detection of cocaine and its two major metabolites, benzoylecgonine and EME (75). Using a double-pass development, first utilizing a relatively hydrophobic development system to move other drugs and interferences well up the chromatogram, and a second more polar system to move the metabolites off the origin, one can sensitively detect cocaine and its two major metabolites. Sensitivity is considerably better if Dragendorf's rather than an iodoplatinate reagent is used for detection. Following recent cocaine use a typical pattern of cocaine and each of its two major metabolites is apparent and the latter may persist for about 24–48 hr. As always, TLC requires confirmation with a more specific procedure such as GC/MS when employed for drug abuse screening, but does offer an alternative or adjunct to immunoassay.

ACKNOWLEDGMENTS

Preparation of this manuscript was supported in part by NIDA Grant 1 P50 DA04060 and NIMH Grant MH30929.

REFERENCES

1. R. Byck Ed., *Cocaine Papers*, Stonehill, New York, 1974.

2. C. Van Dyke and R. Byck, Cocaine, *Sci. Am., 246*, 128–141 (1982).

3. R. Peterson, "Cocaine: An Overview," In R. Petersen and R. Stillman, Eds., *Cocaine: 1977,* National Institute on Drug Abuse Research Monograph 13, DHHS Pub No (ADM) 77-471, U.S. Government Printing Office, Washington DC, 1977.

4. B. Holmstedt and A. Fredga, "Sundry Episodes in the History of Coca and Cocaine," *J. Ethnopharmacol., 3* (2–3), 113–147 (1981).

5. N. J. Kozel and E. H. Adams, Eds., *Cocaine Use in America: Epidemiologic and Clinical Perspectives,* National Institute on Drug Abuse Research Monograph 61, DHHS Pub No (ADM) 85-1414, U.S. Government Printing Office, Washington DC, 1985.

6. "Scientific Perspectives on Cocaine Abuse," sponsored by American Society for Pharmacology and Experimental Therapy and Committee on Problems of Drug Dependence, *Pharmacologist, 29,* 20–27 (1987).

7. L. L. Cregler and H. Mark, "Medical Complications of Cocaine Abuse," *N. Engl. J. Med., 315,* 1495–1500 (1986).

8. M. W. Fischman, "The Behavioral Pharmacology of Cocaine in Humans," in J. Grabowski, Ed., *Cocaine: Pharmacology, Effects and Treatment of Abuse,* NIDA Research Monograph 50, National Institute on Drug Abuse, Rockville, MD, 1984, pp. 72–91.

9. R. T. Jones, "The Pharmacology of Cocaine," in J. Grabowski, Ed., *Cocaine: Pharmacology, Effects and Treatment of Abuse,* NIDA Research Monograph 50, National Institute on Drug Abuse, Rockville, MD, 1984, pp. 34–53.

10. C. Van Dyke, P. G. Barash, P. Jatlow, and R. Byck, "Cocaine: Plasma Concentrations after Intranasal Application in Man," *Science, 191,* 859–861 (1976).

11. C. E. Johanson, "Assessment of the Dependence Potential of Cocaine in Animals," in J. Grabowski, Ed., NIDA Research Monograph 50, National Institute on Drug Abuse, Rockville, MD, 1984, pp. 54–71.

12. M. A. Bozarth and R. A. Wise, "Toxicity Associated with Long-term Intravenous Heroin and Cocaine Self-administration in the Rat," *J. Am. Med. Assoc., 254,* 81–83 (1985).

13. M. W. Fischman, C. R. Schuster, J. Javaid, Y. Hatano, and J. Davis, "Acute Tolerance Development to the Cardiovascular and Subjective Effects of Cocaine," *J. Pharmacol. Exp. Ther., 235,* 677–682 (1985).

14. C. Van Dyke, J. Ungerer, P. Jatlow, P. Barash, and R. Byck, "Intranasal Cocaine: Dose Relationships of Psychological Effects and Plasma Levels," *Intl. J. Psychiatry Med., 11*(4), 391–403 (1982).

15. N. H. G. Holford and L. B. Sheiner, "Understanding the Dose Effect Relationship: Clinical Application of Pharmacokinetic–Pharmacodynamic Models," *Clin. Pharmacokinet., 6,* 429–453 (1981).

16. R. Zahler, P. Wachtel, P. Jatlow, and R. Byck, "Kinetics of Drug Effect by Distributed Lags Analysis: An Application to Cocaine," *Clin. Pharmacol. Ther. 31,* 775–782 (1982).

17. M. J. Chou, J. J. Ambre, and T. I. Ruo, "Kinetics of Cocaine Distribution, Elimination, and Chronotropic Effects," *Clin. Pharmacol. Ther., 38,* 318–324 (1986).

18. C. V. Wetli and D. A. Fishbain, "Cocaine-induced Psychosis and Sudden Death in Recreational Cocaine Wars," *J. Forensic Sci., 30,* 873–880 (1985).

19. J. Isner, N. A. Mark Estes, III, P. D. Thompson, M. R. Costanzo-Nordin, R. Subramanian, G. Miller, G. Katsas, K. Sweeney, and W. Q. Sturner, "Acute Cardiac Events Temporally Related to Cocaine Abuse," *N. Engl. J. Med., 315,* 1438–1443 (1986).

20. R. W. Simpson and W. D. Edwards, "Pathogenesis of Cocaine-induced Ischemic Heart Disease," *Arch. Pathol. Lab. Med., 110,* 479–484 (1986).

21. M. H. Evans, C. Dwivedi, and R. D. Harbison, "Enhancement of Cocaine-

induced Lethality by Phenobarbital," in E. H. Ellinwood and M. M. Kilbey, Eds., *Cocaine and Other Stimulants,* Plenum, New York, 1975, pp. 253–267.

22. M. A. Evans and R. D. Harbison, "Cocaine-induced Hepatotoxicity in Mice," *Toxicol. Appl. Pharmacol., 45,* 739–754 (1978).

23. R. W. Freeman and R. D. Harbison, "Hepatic Periportal Necrosis Induced by Chronic Administration of Cocaine," *Biochem. Pharmacol., 30,* 777–783 (1981).

24. M. N. Kloss, G. M. Rosen, and E. J. Rauckman, "Cocaine Mediated Hepatotoxicity, a Critical Review," *Biochem. Pharmacol., 33,* 169–173 (1984).

25. L. Shuster, F. Quimby, A. Bates, and M. S. Thompson, "Liver Damage from Cocaine in Mice," *Life Sci., 20,* 1035–1042 (1977).

26. A. C. Smith, R. W. Freeman, and R. D. Harbison, "Ethanol Enhancement of Cocaine-induced Hepatotoxicity," *Biochem. Pharmacol., 30*(5), 453–458 (1981).

27. M. L. Thompson, L. Shuster, and K. Shaw, "Cocaine-induced Hepatic Necrosis in Mice: The Role of Cocaine Metabolism," *Biochem. Pharmacol., 28,* 2839–2395 (1979).

28. J. Itkonen, S. Schnoll, and J. Glassoroth, "Pulmonary Dysfunction in "Freebase" Cocaine Users," *Arch. Intern. Med., 44,* 2195–2197 (1984).

29. R. D. Weiss, P. D. Goldenheim, S. M. Mirin, C. A. Hales, and J. H. Mendelson, "Pulmonary Dysfunction in Cocaine Smokers," *Am. J. Psychiatry, 138,* 1110–1112 (1981).

30. C. Van Dyke, P. Jatlow, J. Ungerer, P. G. Barash, and R. Byck, "Oral Cocaine: Plasma Concentrations and Central Effects," *Science, 200,* 211–213 (1978).

31. P. Wilkinson, C. Van Dyke, P. Jatlow, P. Barash, and R. Byck, "Intranasal and Oral Cocaine Kinetics," *Clin. Pharm. Ther., 27,* 386–394 (1980).

32. B. Holmstedt, J. E. Lindgren, L. Rivier, and T. Plowman, "Cocaine in Blood of Coca Chewers," *J. Ethnopharmacol. 1,* 69–78 (1979).

33. D. Paly, C. Van Dyke, P. Jatlow, F. Cabieses, R. Byck, "Cocaine Plasma Concentrations in Coca Chewers," *Clin. Pharmacol. Ther., 25,* 240 (1979), abstract.

34. G. Barnett, R. Hawks, and R. Resnick, "Cocaine Pharmacokinetics in Humans," *J. Ethnopharm., 3,* 353–366 (1981).

35. M. J. Kogan, K. G. Vereby, A. C. dePace, R. B. Resnick, and S. J. Mule, "Quantitative Determination of Benzoylecgonine and Cocaine in Human Biofluids by Gas Liquid Chromatography," *Anal. Chem., 49,* 1965–1969 (1977).

36. M. W. Fischman, C. R. Schuster, L. Resnekov, F. E. Shick, N. A. Krasnegor, W. Fennell, D. X. Freedman, "Cardiovascular and Subjective Effects on Intravenous Cocaine Administration in Humans," *Arch. Gen. Psychiatry, 33,* 983–989 (1976).

37. J. I. Javaid, M. W. Fischman, C. R. Schuster, H. Dekirmenjiian, and J. M.

Davis, "Cocaine Plasma Concentrations: Relation to Physiological and Subjective Effects in Humans," *Science, 200,* 227–228 (1978).

38. D. Paly, P. Jatlow, C. Van Dyke, R. Jeri, and R. Byck, "Plasma Cocaine Concentrations during Cocaine Paste Smoking," *Life Sci., 30*(9), 731–738 (1982).

39. M. Perez-Reyes, S. Diguiseppi, G. Ondrusek, A. R. Jeffcoat, and C. E. Cook, "Free-base Cocaine Smoking," *Clin. Pharmacol. Ther., 32,* 459–465 (1982).

40. F. R. Jeri, C. Sanchez, T. Del Pozo, and M. Fernandez, "The Syndrome of Coca Paste," *J. Psychedelic Drugs, 10,* 361–370 (1978).

41. F. H. Gawin and H. D. Kleber, "Cocaine Abuse in a Treatment Population: Patterns and Diagnostic Distinctions," in E. H. Adams and N. J. Kozel, Eds., *Cocaine Use in America: Epidemiologic and Clinical Perspectives,* NIDA Monograph Series 61, National Institute on Drug Abuse, Rockville, MD, 1985, pp. 182–192.

42. R. K. Siegel, "Cocaine Smoking," *J. Psychoactive Drugs, 14,* 277–359 (1982).

43. F. Fish and W. D. C. Wilson, "Excretion of Cocaine and its Metabolites in Man," *J. Pharm. Pharmacol., 21,* 1355–1385 (1969).

44. J. Ambre, M. Fischman, and T. I. Ruo, "Urinary Excretion of Ecgonine Methyl Ester, a Major Metabolite of Cocaine in Humans," *J. Anal. Toxicol., 8,* 23–25 (1984).

45. J. J. Ambre, J. H. Ruo, G. L. Smith, D. Backer, and C. M. Smith, "Ecgonine Methyl Ester, a Major Metabolite of Cocaine," *J. Anal. Toxicol., 6,* 26–29 (1982).

46. J. Ambre, "The Urinary Excretion of Cocaine and Metabolites in Humans: A Kinetic Analysis of Published Data," *J. Anal. Toxicol., 9,* 241–245 (1985).

47. R. L. Hawks, I. J. Kopin, R. W. Colburn, and N. B. Thoa, "Nor-cocaine: A pharmacologically Active Metabolite of Cocaine Found in Brain," *Life Sci., 15,* 2189–2195 (1974).

48. T. Inaba, D. J. Stewart, and W. Kalow, "Metabolism of Cocaine in Man," *Clin. Pharmacol. Ther., 23,* 547–552 (1978).

49. P. Jatlow, P. G. Barash, C. Van Dyke, J. Radding, and R. Byck, "Cocaine and Succinylcholine Sensitivity: A New Caution," *Anesth. Analg., 58,* 235–238 (1979).

50. M. Mayersohn and D. Perrier, "Kinetics of Pharmacologic Response to Cocaine," *Res. Commun. Chem. Pathol. Pharmacol., 22,* 465–474 (1978).

51. R. M. Smith, M. A. Poquette, and P. J. Smith, "Hydroxymethoxybenzoylmethylecgonines: New Metabolites of Cocaine from Human Urine," *J. Anal. Toxicol., 8,* 29–34 (1984).

52. R. M. Smith, "Arylhydroxy Metabolites of Cocaine in the Urine of Cocaine Users," *J. Anal. Toxicol., 8,* 35–37 (1984).

53. D. J. Stewart, T. Inaba, B. K. Tang, W. Kaklow, "Hydrolysis of Cocaine in Human Plasma by Cholinesterase," *Life Sci., 20,* 1557–1564 (1977).

54. D. J. Stewart, T. Inaba, M. Lucassen, and W. Kalow, "Cocaine Metabolism: Cocaine and Norcocaine Hydrolysis by Liver and Serum Esterases," *Clin. Pharmacol. Ther., 25,* 464–468 (1979).

55. G. Leighty and A. F. Fentiman, "Metabolism of Cocaine to Norcocaine and Benzoylecgonine by an in Vitro Microsomal Enzyme System," *Res. Commun. Pathol. Pharmacol., 8,* 65–74 (1974).

56. R. M. Smith, "Ethyl esters of arylhydroxy and arylhydronymethoxy cocaines in the urines of simultaneous cocaine and ethanol users," *J. Anal. Toxicol., 8,* 38–42 (1984).

57. C. V. Wetli and R. K. Wright, "Death Caused by Recreational Cocaine Use," *J. Am. Med. Assoc., 241,* 2519–2522 (1981).

58. R. E. Mittleman and C. V. Wetli, "Death Caused by Recreational Cocaine Use, an Update," *J. Am. Med. Assoc., 252,* 1889–1893 (1984).

59. M. S. Gold, A. M. Washton, and C. A. Dackis, "Cocaine Abuse: Neurochemistry, Phenomenology and Treatment," in N. J. Kozel and E. H. Adams, Eds., *Cocaine Use in America: Epidemiological and Clinical Perspectives,* NIDA Monograph Series No. 61, U.S. Government Printing Office, Washington DC, 1985, pp. 130–157.

60. R. A. Wise, "Neural Mechanisms of the Reinforcing Action of Cocaine," in J. Grabowski, Ed., *Cocaine: Pharmacology, Effects, and Treatment of Abuse,* NIDA Monograph Series No. 50, Washington, DC, U.S. Government Printing Office, 1984, pp. 15–33.

61. R. M. Post, S. R. B. Weiss, A. G. U. Pert, et al., "Chronic Cocaine Administration: Sensitization and Kindling Effects," in S. Fisher, A. Raskin and E. H. Uhlenmuth, Eds., *Cocaine: Clinical and Biobehavioral Aspects,* Oxford University Press, New York, 1987, pp. 109–173.

62. M. E. A. Reith, D. L. Allen, H. Sershen, and A. Lajtha, "Similarities and Differences between High-affinity Binding Sites for Cocaine and Imipramine in Mouse Cerebral Cortex," *J. Neurochem., 43,* 249–255 (1984).

63. M. E. A. Reith, H. Sershen, D. L. Allen, and A. Lajtha, "A portion of (^{3}H)Cocaine Binding in Brain is Associated with Serotonergic Neurons," *Mol. Pharmacol., 23,* 600–606 (1983).

64. P. Jatlow, and D. Bailey, "Gas Chromatographic Analysis for Cocaine in Human Plasma with Use of a Nitrogen Detector," *Clin. Chem., 21,* 1918–1921 (1975).

65. P. Jacobs, B. A. Elias-Baker, R. T. Jones, and N. L. Benowitz, "Determination of Cocaine in Plasma by Automated Gas Chromatography," *J. Chromatogr., 306,* 173–181 (1984).

66. J. I. Javaid, H. Dekirmenjian, J. M. Davis, and C. R. Schuster, "Determination of cocaine in human urine, plasma and red blood cells by gas–liquid chromatography," *J. Chromatogr., 152,* 105–113 (1978).

67. J. E. Wallace, H. E. Hamilton, D. E. King, et al., "Gas–Liquid Chromatographic Determination of Cocaine and Benzoylecgonine in Urine," *Anal. Chem., 48,* 34–38 (1976).

68. P. Jatlow, C. Van Dyke, P. Barash, and R. Byck, "Measurement of Cocaine and Benzylecgonine in Urine and Separation of Various Cocaine "Metabolites" Using Reverse Phase High Performance Liquid Chromatography," *J. Chromatogr., 152,* 115–121 (1978).

69. M. A. Evans and T. Morarity, "Analysis of Cocaine and Cocaine Metabolites by High Pressure Liquid Chromatography," *J. Anal. Toxicol., 4,* 19–20 (1980).

70. R. L. Foltz, A. F. Fentiman, and R. B. Foltz, Eds. *GC/MS Assays for Abused Drugs in Body Fluid,* NIDA Research Monograph 32, U.S. Government Printing Office, Washington, DC, 1980.

71. D. M. Chinn, D. J. Crouch, M. A. Peat, B. S. Finkle, and T. A. Jennison, "Gas Chromatography–Chemical Ionization Mass Spectrometry of Cocaine and Its Metabolites in Biological Fluids," *J. Anal. Toxicol., 4*(1), 37–42 (1980).

72. S. P. Jindahl, T. Lutz, and P. Vestergaard, "Mass Spetrometric Determination of Cocaine and its Biologically Active Metabolite, Norcocaine in Human Urine," *Biomed. Mass. Spectrom., 5,* 658–662 (1978).

73. S. J. Mule, M. L. Bastos, and D. Jukofsky, "Evaluation of Immunoassay Methods for Detection in Urine of Drugs Subject to Abuse," *Clin. Chem., 20,* 243 (1974).

74. S. J. Mule, D. Jikovsky, M. Kogan, et al., "Evaluation of RIA for Benzoylecgonine (a Cocaine Metabolite) in Human Urine," *Clin. Chem., 23,* 796–801 (1977).

75. J. E. Wallace, H. E. Hamilton, H. Schwertner, et al. "Thin-layer Chromatographic Analysis of Cocaine and Benzoylecgonine in Urine," *J. Chromatogr., 114,* 433–441 (1975).

CHAPTER

10

GAS CHROMATOGRAPHY/MASS SPECTROMETRY SCREENING FOR ANABOLIC STEROIDS

R. H. BARRY SAMPLE AND JOHN C. BAENZIGER

Indiana University Medical Center
Indianapolis, Indiana

1. INTRODUCTION

Synthetic anabolic steroids have been in use in clinical medicine since their development in the 1940s (1). However, their analysis did not become important until the 1970s, when methods for their detection were developed as part of doping control programs in athletics. The detection of anabolic steroids came to international view during the 1976 Summer Olympics in Montreal and ever since the analysis of these drugs has been of great importance. A summary of anabolic steroids including their use, metabolism, and analytical identification is included. The discussion that follows relates mostly to the use of various analytical techniques for broad-spectrum screening of anabolic steroids in biological fluids. A selected method is provided that allows the detection for most of the anabolic steroids of interest in athletic doping control. This method has been extensively used in our laboratory for several years and was the basic procedure used for anabolic steroid screening for the Tenth Pan American Games, in 1987.

2. CLINICAL IMPORTANCE

"Doping" is a term for the use of substances by athletes for the sole purpose of increasing their athletic performance in an unfair or artificial manner. The use of anabolic steroids as doping agents in athletics began in the 1950s and has steadily increased to the present day (1, 2). These drugs are used by athletes to improve the development of muscle mass, strength, and power, and in some cases muscle definition (2). There has been significant debate in the scientific literature over whether these compounds actually enhance performance (1–3). It is now felt that when they are used in the proper setting, there is a positive effect on performance (1, 2). This effect is found when these drugs are used by a trained athlete with a structured diet and exercise program. The use of these drugs has been largely uncontrolled by medical practice and the athletes commonly use supraphysiological doses (10–40 times the recommended doses) during training periods (2). The primary concern of medical personnel involved with athletes using these drugs are the risks of complications from anabolic steroids and the lack of information on the long-term effect of these drugs (2). Table 10.1 lists the complications of anabolic steroid use reported in literature reviews (1, 2, 4–6). Recent reports, since the reviews of Haupt and Rovere (1) and Lamb (2) have investigated the association between altered cholesterol metabolism and the possibility of accelerated atherosclerosis resulting from the long-term use of anabolic steroids (2, 5, 7, 8). Owing to these medical risks and to prevent unfair competition, many athletic organizations such as the International Olympic Committee, International Amateur Athletic Federation, and the National Collegiate Athletic Association have banned the use of steroids by athletes.

Table 10.1. Complications Associated With Anabolic Steroid Use

Liver Dysfunction and Disorders
Increased AST, ALT, LDH enzymes
Peliosis hepatis
Hepatocellular carcinoma

Lipoprotein Abnormalities
Increased total cholesterol
Increased LDL cholesterol
Decreased HDL cholesterol

Endocrine Organ Dysfunction
Decreased testosterone, FSH, LH
Testicular atrophy
Menstrual irregularity
Virilization in women and youths

Other Abnormalities
Premature epiphyseal closure
Aggressive behavior

Source: Information from refs. 1, 2, 4–8.

3.   CHEMICAL STRUCTURE AND METABOLISM

3.1.   Chemical Structure

The synthetic anabolic steroids are structurally related to the native testosterone molecule. They were designed to maintain the anabolic effect of testosterone while reducing the androgenic effect. As can be seen from the structures of the commonly available anabolic steroids shown in Fig. 10.1 the presence of the 17β hydroxyl group is common to all steroids with anabolic function. The various synthetic anabolic steroids differ in the presence of substitutions at the 17α hydrogen and by modifications of the A-ring of the testosterone steroid nucleus. With the development of methodologies for anabolic steroid detection, there was an increased use of exogenous testosterone by athletes. The detection of exogenous testosterone use has relied on the increase in the ratio of testosterone to epitestosterone when testosterone is administered exogenously. A ratio of >6:1 is considered abnormal by various international athletic federations. The chemical structure of the various synthetic anabolic steroids

CHLOROTESTOSTERONE

FLUOXYMESTERONE

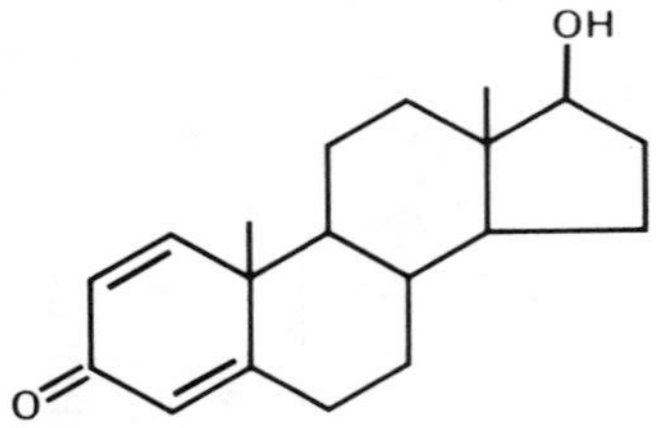

METHANDIENONE

BOLDENONE

ETHYLESTRENOL

METHANDRIOL

BOLASTERONE

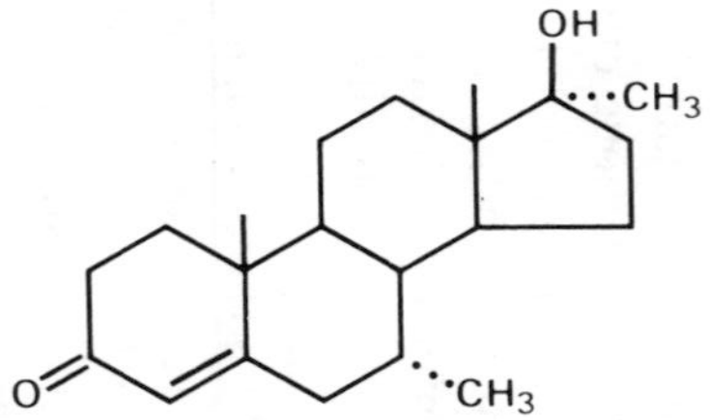

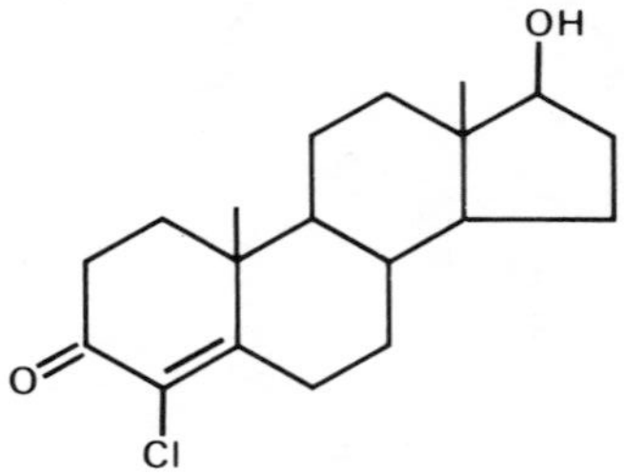

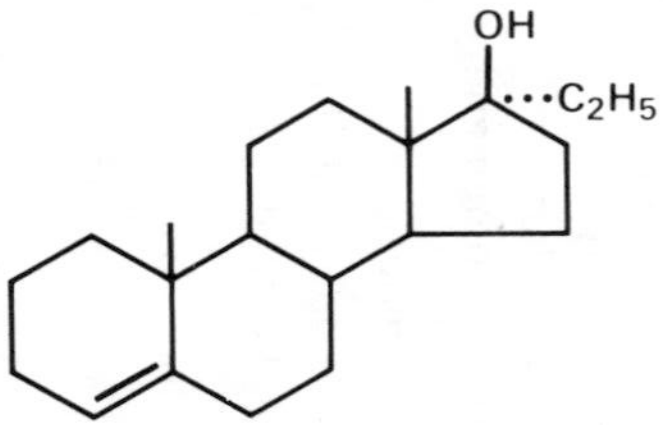

DEHYDROCHLOROMETHYL
TESTOSTERONE

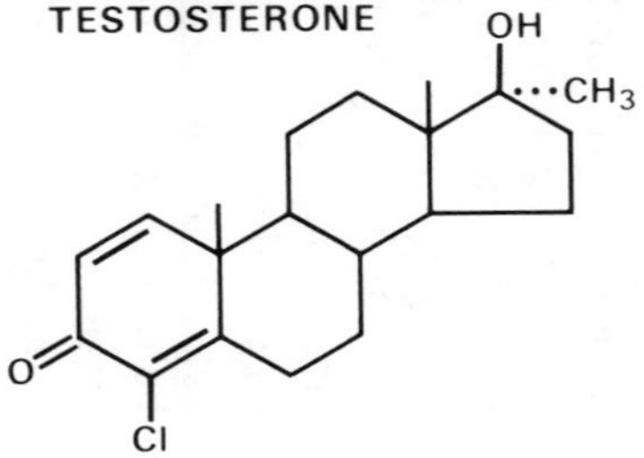

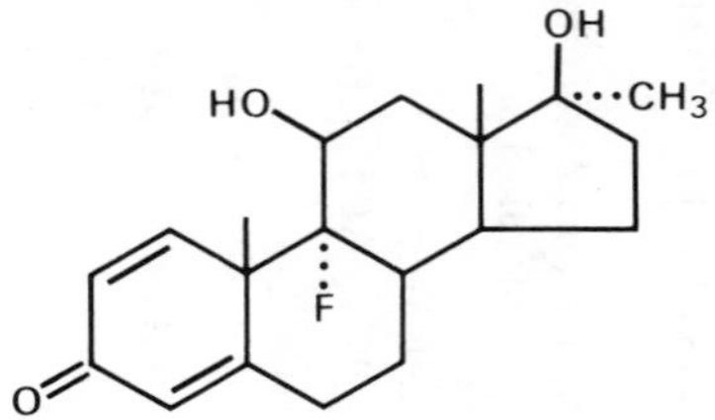

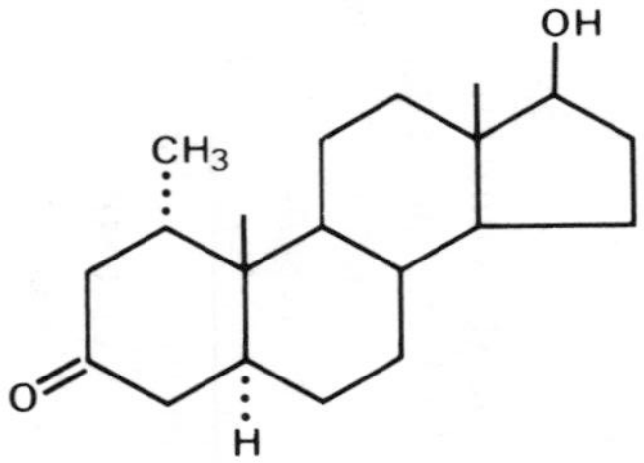

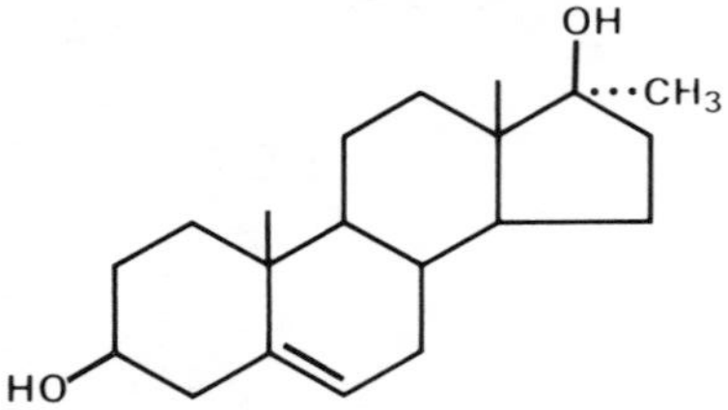

MESTEROLONE

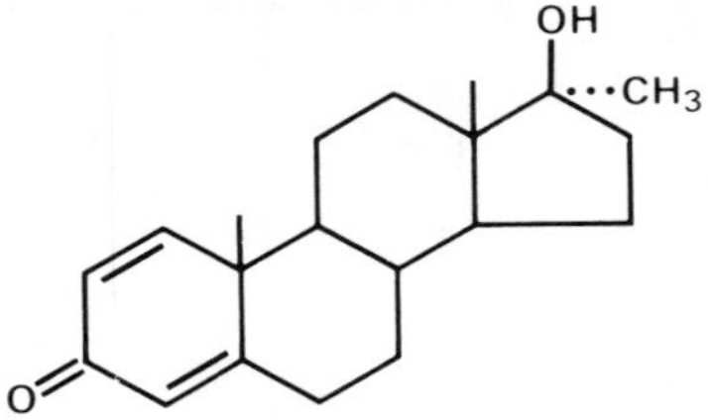

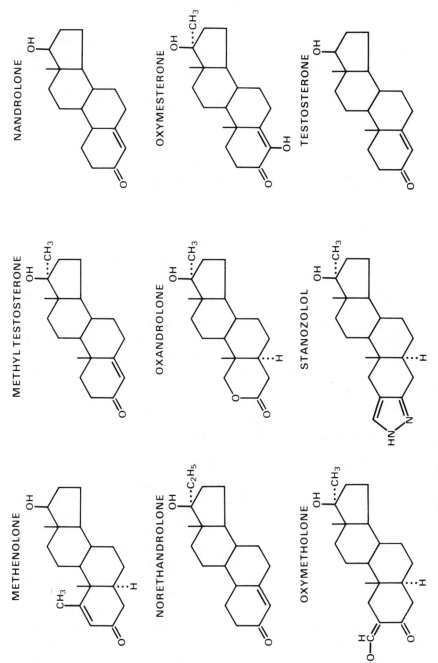

Figure 10.1. Structure of anabolic steroids commonly used by athletes.

251

can be linked to the amount of anabolic versus androgenic activity the compound exhibits (9). Athletes may commonly use several different compounds simultaneously or in a sequential fashion to maximize the effect and minimize the side effects of each drug. The structure of the anabolic steroids is also important in relation to the metabolic fate of the drug within the body.

3.2. Metabolism

Many synthetic steroids are extensively metabolized, and their metabolism follows pathways similar to that of testosterone. Testosterone metabolism as shown in Fig. 10.2 (10, 11) proceeds by two pathways: (1) oxidation of the 17-hydroxyl to the 17-keto function and (2) reduction of the A-ring. The resulting major metabolites are the 3-hydroxyl-17-keto isomers, androsterone, and etiocholanolone. Testosterone is also metabolized to androstenedione, which can be transformed into estrogens (10). Finally, some of the metabolites are eliminated as glucuronide or sulfate conjugates.

The metabolism of synthetic steroid compounds that lack alkyl substitutions at the 17α carbon results in 17-keto metabolites (12–19) similar to testosterone. The 17-alkyl-substituted steroids do not result in appeciable levels of a 17-keto metabolite (16, 19–29). In both 17α-substituted and nonsubstituted steroids there are commonly changes in the A-ring during metabolism. A-ring reduction of the 3-keto group and the saturation of the C-4–C-5 double bond yield androst-3-ol compounds with isomers resulting from the formation of stereo centers at the C-3 and C-5 positions. Of importance is the preference of the body to form 3α hydroxyl and 5β H isomers (10, 15, 17, 24). The presence of an androst-1,4-diene-3-one structure stabilizes A-ring reductions (17, 22, 27), and only selected compounds exhibit any significant reduction of this ring (17). A metabolic pathway involving 6β hydroxylation is seen in methandienone (methandrostenolone) and boldenone, both of which have the stabilized A-ring structure (22, 24, 27). This pathway has also been reported with prednisone, prednisolone, and cortisol (17, 24). The formation of 16-hydroxyl metabolites has been reported for oxandrolone, dehydrochloromethyltestosterone, and testosterone (11, 21, 27). The epimerization of the 17-hydroxyl function has been described with methandienone (20, 22).

The metabolism of 17-ethyl substituted 19-nortestosterone analogs, such as norethandrolone and ethylestrenol, involves both A-ring reduction and hydroxylation of the ethyl side chain (12, 30), yielding a 19-norpregnantriol metabolite. Many of the synthetic anabolic steroids or metabolites are conjugated with glucuronide after metabolism and thus

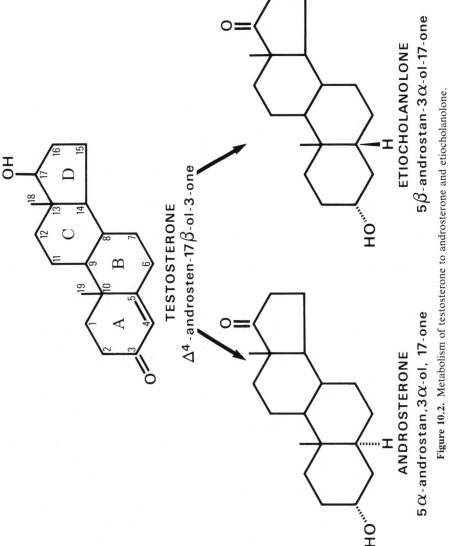

TESTOSTERONE
Δ^4-androsten-17β-ol-3-one

ANDROSTERONE
5α-androstan,3α-ol, 17-one

ETIOCHOLANOLONE
5β-androstan-3α-ol-17-one

Figure 10.2. Metabolism of testosterone to androsterone and etiocholanolone.

253

are present in the conjugated fraction. Only a few studies have discussed the actual pharmacokinetics of steroid elimination (21, 29, 31), but many of the steroids can be detected for weeks and months after their use.

4. ANALYTICAL METHODS

4.1. Radioimmunoassay

Radioimmunoassay in steroid analysis has been employed for many years. The development of antibodies directed toward specific anabolic steroids has been used in the analysis of cattle feeds (32). In the late 1970s Brooks (33) developed several antibodies that were directed toward the detection of three groups of steroids: (1) 19-nortestosterone analogs; (2) norethandrolone analogs; and (3) 17α-substituted analogs. These antibodies had a broad specificity range and would react with many of the then common anabolic steroids. Subsequently the antibodies were combined to yield two RIA kits for the detection in urine of 19-nortestosterone/norethandrolone analogs and 17α-substituted steroids.

The RIA kits developed by Brooks require an initial organic extraction of the steroids from urine, followed by an immunoassay procedure. Iodine-125 is used as the tracer; separation is performed by charcoal adsorption, with the bound fraction counted. Although sensitive for many of the anabolic steroids of interest in athletics these assays suffer from having insufficient cross-reactivity for detection of several anabolics, such as methenolone and stanozolol (14, 31). Table 10.2 shows the cross-reactivity of these kits (31), along with antibodies developed by Hampl and Starka (34), toward a variety of anabolic steroids and normal urinary steroids. The cross-reactivity results are expressed in percentages as compared to the assay reactivity of the standards methandienone (17-methyl), 19-nortestosterone (19-nor), and norethandrolone (17-ethyl). The cross-reactivity for several anabolic steroids is low as compared to the standards; therefore the RIA procedure may fail to detect the presence of these steroids. For example, the 17-methyl assay of Brooks is 25 times less sensitive for the detection of stanozolol as compared to methandienone (~4% vs. 100%) and may not even detect the presence of methenolone (0.1%). The cross-reactivity with normal urinary steroids, such as testosterone, and components of birth control medications (norethindrone and norgestrel) should be noted. If testosterone or norethindrone is present in sufficient quantities then the screen can result in a false positive, owing to testosterone's cross-reaction of ~9% and norethindrone's cross-reaction of 65% in the 17-methyl and 19-nor RIA assays of

Table 10.2. RIA Cross-Reactivity of Synthetic Anabolic and Normal Urinary Steroids Expressed as Percent of Assay Reactivity of Standard

Compound	Dugal (31)			Hampl and Starka (34)
	19-Nor (%)	17-Ethyl (%)	17-Methyl (%)	17-Methyl (%)
Methandienone	<1.3	0.6	100.0	100.0
17-Methyltestosterone		3.7	169.7	163.0
Dimethylandrostanolone				39.5
Chlorotestosterone			76.9	39.5
Fluoxymesterone		<0.4	71.8	6.5
Mestanolone			34.5	
Oxandrolone			20.0	
Stanozolol			3.6	7.2
Oxymetholone			1.1	9.6
Drostanolone				0.4
Methenolone				0.1
Methandrostenediol			<0.7	39.0
Norethandrolone	113.0	100	<0.7	5.5
Ethylestrenol	<1.3	28.0	<0.7	
19-Nortestosterone	100.0	<0.4	0.6	1.9
Testosterone	0	<0.4	9.0	19.8
Testosterone acetate			0	0.02
Dihydrotestosterone				9.7
Norethindrone	65.0	6.6	~5.0	
Norgestrel	75.0	<0.4	~5.0	

Brooks, respectively. The cross-reaction with testosterone can be eliminated if the specimen is acetylated prior to the analysis, but this requires an additional step. To avoid false positive results from the RIA screening, other analytical methods must be used to identify definitively the presence of a synthetic anabolic steroid in a urine found positive by this screening procedure (31). With the recent increased use of exogenous testosterone by athletes, it has also become necessary to measure both testosterone and epitestosterone on each urine sample, which would require two additional RIA procedures.

RIA was used to screen for anabolic steroids in urines during the 1976 Summer Olympics in Montreal, with positive screens being confirmed by GC/MS technology (31). During these games eight of 275 athletes tested were found positive by this technique. Only a few articles in the literature reference the use of these assays for doping control programs (31, 34–

37). It should be noted that the cross-reactivity studies that have been reported do not include studies on the major metabolites of the synthetic anabolic steroids. Owing to the problems with both sensitivity and specificity of these assays, and the improvement in automated systems for GC/MS, most doping control procedures for anabolic steroids in athletics have relied on GC/MS for screening.

4.2. Thin-Layer Chromatography

The use of thin-layer chromatography (TLC) methods for the analysis of anabolic steroids has been primarily of value in research studies of steroid metabolism (14, 17, 18, 21, 24–29) and in veterinary science (38–40). TLC has been used as the primary screening method in the veterinary field (40) for investigating samples from meat and animal feeds for the presence of anabolic agents. These procedures use silica plates with one- or two-dimensional migration followed by acid/alcohol-induced fluorescence for detection. Verkbeke (40) reported a procedure for the detection of several parent anabolic steroids in muscle tissue and urine of animals. After organic and solid-phase extraction of the sample, the residues were spotted on silica plates and chromatographed in two dimensions with chloroform/ethanol/benzene (36:1:4) and hexane/dichloromethane/ethylacetate (1:2:2) solvents. The acid-induced fluorescence was performed using 5% sulfuric acid/ethanol spraying and illumination at 366 nm. The on-plate detection limits were 1–5 ng of applied steroids. The procedure described by Verkbeke allowed for 10 analyses per week by one analyst, requiring 1.5 days for the analysis of five samples for both estrogen and androgen compounds.

The use of TLC analysis of steroids in athletes has not been reported. The main problems with TLC analysis for screening in athletics have been its sensitivity and specificity (14).

4.3. High Pressure Liquid Chromatography

High pressure liquid chromatography has been used for the separation and identification of anabolic steroids mostly in the veterinary field (41–46) and in the pharmaceutical industry (47–49). Only a few reports discuss the use of HPLC for anabolics of interest to humans or for screening purposes (44, 45, 50). The detection in urine of methandienone has been reported by Frischkorn (45) with a detection limit of 1 ng applied on-column. Reverse phase bonded (C_{18}) columns have been the predominant column used; however, diol phase columns (44) and silica (43) columns have also been reported. UV detection has been the primary type used in these studies, but improved sensitivity has been achieved with fluorescence detection of dansyl derivatives (41), as well as chemiluminescence (46). No extensive studies of the use of HPLC analyses for urine screening have been reported in the literature. LC/MS, with its enhanced specificity, may be of some future use in the area of doping control for the analysis of glucocorticoid type steroids.

4.4. Gas Chromatography/Mass Spectroscopy

Several reviews have been written on the use of gas chromatographic techniques with mass spectroscopy for the analysis of steroids in general (51–54). These reviews discuss in detail the analytical factors involved in steroid analysis. A short summary that discusses the hydrolysis, extraction, and derivatization of samples for anabolic steroid analysis is presented in this section.

4.4.1. Extraction

Historically steroids have been extracted using solvents such as ethyl acetate, methylene chloride, diethyl ether, or chloroform–alcohol mixtures. These procedures were often time consuming, had low recoveries requiring large volumes of urine and solvent, were prone to emulsion formation, and were not very reproducible, which presented problems in quantitative analysis. Furthermore, different solvent systems were required for steroids of differing polarity.

The use of solids such as XAD-2 resin (55) greatly improved the extraction of both conjugated and nonconjugated steroids. However, not all polar steroids conjugates, especially the sulfoconjugates, are well extracted unless the column is prewashed with triethylamine (56) or sodium sulfate before elution with methanol. The XAD-2 columns also require extensive washing to remove contaminants and flow rates are relatively low (~0.2 mL/min). Sephadex type gels (Lipidex-1000, Sephadex-LH-20) have been widely used for steroid extraction. By selecting the appropriate stationary phase, it is possible quantitatively to extract selected steroids with common functional groups. Lipidex-1000, one of the common gels, is suitable for steroids of low to medium polarity (57). However, steroid conjugates are not well retained; thus hydrolysis is required prior to extraction. The Sephadex-LH-20 gels are lipophilic cation exchangers that are often used for subfractionation of steroid extracts (57–59).

With the development of the small disposable bonded phase columns (C_{18}), steroid extraction was greatly improved. Although the C_{18} moiety is similar to the alkyl substitution of the Lipidex, these columns extract steroids with a wider polarity range than either solvent extraction, XAD-2, or Lipidex (54, 57–60). In addition, the columns have a very high capacity (>100 ml) and flow rates (15–30 ml/min) (60). The C_{18} columns provide a method for the rapid extraction of a wide variety of steroids (both free and conjugated) including all banned anabolic steroids without the need for further chromatographic purification.

4.4.2. Hydrolysis

Following the initial extraction of free and conjugated steroids from the urine, it is necessary to hydrolyze the extract because many of the anabolic steroids are excreted as glucuronide and sulfate conjugates. This hydrolysis is usually performed with the enzyme β-glucuronidase/arylsulfatase from *Helix pomatia*. This hydrolysis has the advantage that the total endogenous urinary steroid profile may be assessed. However, a significant disadvantage is that this preparation may contain an isomerase that can convert androstenediol to testosterone. Therefore, in order to determine testosterone levels accurately, it is necessary to use β-glucuronidase from *E. coli*. The *E. coli*-derived enzyme has a shorter incubation time (1 hr at 60°C for *E. coli,* versus 3 hr at 60°C or 16 hr at 37°C for *Helix pomatia*); however, because the *E. coli* preparation does not have arylsulfatase activity, some anabolic and many endogenous steroid metabolite conjugates are not hydrolyzed.

4.4.3. Derivatization

In order to improve their chromatographic and mass spectral characteristics, steroids are usually derivatized before injection on the gas chromatograph. Much of the early work in steroid analysis was done using methoxime (MO)–trimethylsilyl (TMS) derivatives, because they seemed to form the best derivatives for analysis of a wide variety of steroids (61). The methoxime derivatization was used to protect keto functions from enolization with the TMS reagents. When only TMS derivatives are formed, partial enolization of the keto group may occur depending on the reaction conditions and the silylation reagents used. One of the advantages of the MO–TMS derivatives is that the different reaction times required to derivatize hindered keto and hydroxy groups completely can be used to help elucidate the structure of an unknown steroid. However, a disadvantage is the formation of syn/anti (54) isomers, which will chromatograph at different retention times on the commonly used nonpolar stationary phases. Additionally, these derivatives are acid labile (62) and often require purification on Lipidex or Sephadex gels prior to injection on the gas chromatograph (63, 64).

If only TMS reagents are used, TMS ethers of hydroxyls are usually readily formed. There are many published procedures for the analysis of steroids using TMS ether formation, including the procedure described by Bertrand, Masse, and Dugal (65) which was used during the 1976 Olympics. Depending on the reaction time, silylating strength, and steric hindrance, keto functions may also be derivatized into TMS enols. If the enolization is incomplete, the sensitivity and quantitative precision of the assay are reduced. An advantage of TMS enol formation is improved sensitivity and specificity, because keto–enols may produce very abundant high mass ions in the mass spectrometer. Several catalysts have been used to promote quantitative enol formation: potassium acetate (66), trimethylbromosilane (TMBRS) (67), and trimethyliodosilane (TMIS) (68). Donike (68) compared these different reagents, including TMCS, as catalysts for the TMS derivatization of the 17-hydroxy group in testosterone and found that TMIS was the most efficient catalyst. It was also shown that the bis-silyl (17-OH, 3-keto) testosterone derivative using TMS, triethylsilane, or *tert*-butyldimethylsilane and the appropriate iodinated catalyst produced significantly more abundant molecular ions than the monosilyl derivative or the 3-MO-TMS-derivative. The larger yield of M^+ molecular ion increases sensitivity and is preferable especially for the selected ion monitoring (SIM) quantitation of steroids.

The use of this selective derivatization of steroids with different silylation catalysts can aid in the confirmation of suspected keto-steroids in an unknown sample. It is possible initially to derivatize a sample using a milder catalyst (TMCS) in order to form only the *O*-TMS ether of the suspect compound. After injection on the GC/MS, the sample can then be further derivatized by adding the more potent catalyst, TMIS, to form the bis-TMS (*O*-TMS ether and *O*-TMS keto–enol) derivative. When this sample is reinjected on the GC/MS, the compound should exhibit a shift in retention time and change in ions (Figs. 10.3–10.5) indicating the presence of a steroid with a keto function. This two-step derivatization may result in some loss in sensitivity, because the first step will not produce the abundant high mass ions as seen in the second step.

5. SELECTED ANALYTICAL METHOD:
GAS CHROMATOGRAPHY/MASS SPECTROSCOPY

There have been many procedures published for the GC and GC/MS detection of anabolic steroids in urine, including the analysis of fluoxy-

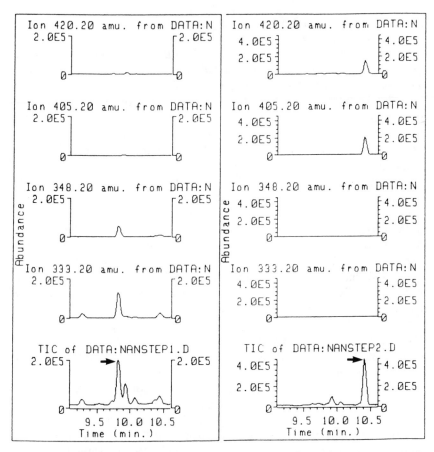

Figure 10.3. Extracted ion current profiles of nandrolone metabolic study (24 hr post-dose). (Left) Nandrolone metabolite, *cis*-19-norandrosterone, derivatized first with 100 μL MSTFA/TMCS (100:2) to form the TMS-hydroxy ether derivative (9.8 min, 348.2 and 333.2 *m/z*). (Right) Same extract after the addition of 100 μL of MSTFA/TMIS (1000:2) to form the TMS-hydroxy ether, TMS-keto-enol derivative (10.4 min, 420.2 and 405.2 *m/z*). Note the increase in the abundance of the bis-TMS derivative (4.0E5) versus the mono-TMS derivative (2.0E5). (See Fig. 10.4 for the mass spectrum of the mono-TMS derivative and Fig. 10.5 for the spectrum of the bis-TMS derivative.)

mesterone (65, 69), methenolone (13), methandienone (19, 21, 22, 65, 69, 70), nandrolone (13, 22, 69, 71), norethandrolone (12, 22), oxandrolone (20, 65), and stanozolol (65, 72). Many of these methods discussed the analysis of individual anabolics, required extensive sample preparation and overnight hydrolysis or derivatization, had inadequate sensitivity or

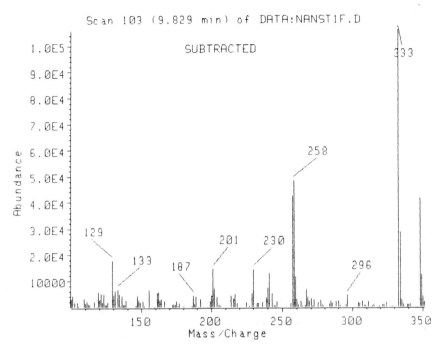

Figure 10.4. Mass spectrum of *cis*-19-norandrosterone TMS ether.

specificity, and therefore were not well suited for the screening of the wide variety of anabolic steroids banned in amateur athletics.

There are several published methods for the simultaneous determination of the different anabolics banned at sporting events (63, 73–76). The method described below can be routinely used for the detection of more than 19 anabolic steroids and metabolites with the simultaneous monitoring of testosterone/epitestosterone ratios (see Table 10.3). This method is slightly modified from that described by Donike (73, 77).

As described in Section 3.2 many of the anabolic and endogenous steroids are excreted as glucuronide or sulfate conjugates. Therefore it is possible to separate the steroids into free and conjugated fractions. Because the free fraction contains mostly exogenously administered steroids but very few endogenous compounds, the detection sensitivity for anabolic steroids is high. In order to have adequate sensitivity (<10 ng/mL) for anabolic steroid screening, it is necessary to use selected ion monitoring (SIM). Catlin (75) used a combined fraction anabolic steroid screen with two ions per compound being monitored. With separation of steroid fractions, the use of up to four or five ions per compound can be monitored

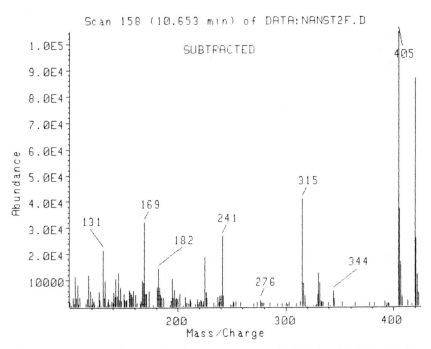

Figure 10.5. Mass spectrum of *cis*-19-norandrosterone TMS Ether–TMS keto Enol.

by SIM if at least 10 SIM groups of ions are used. By separating free from conjugated fractions, both analyses allow for sensitive and specific detection of anabolic steroids.

The method presented here takes advantage of the extraction efficiency of bonded phase C_{18} columns, the use of enzymatic hydrolysis, and the complete silylation of all hydroxyl and keto functions with MSTFA/TMIS in the conjugated fraction assay. The use of capillary columns is essential for the chromatographic separation of the various endogenous and possible anabolic steroids that may be seen in the urine as well as for the sensitivity of detection. Electron impact mass spectroscopy using the SIM mode is the most sensitive detection mode for screening. The availability of multiple ion detection in multiple time separated SIM groups is essential. Using this procedure 20–30 samples can be processed per day by an analyst, with a total analysis time of 12–24 hr.

5.1. Free Fraction Procedure

1. To 5 mL urine add 50 μL internal standard (stanozolol, 10 μg/mL).
2. Extract the urine with 5 mL diethyl ether.

Table 10.3. Anabolic Steroids Detected in Free/Conjugated Fractions[a]

Conjugated Fraction	Free Fraction
Bolasterone	Dehydrochloromethyltestosterone
Boldenone	Fluoxymesterone
Chlorotestosterone	Methandienone
Dromostanolone	Oxandrolone
Ethylestrenol	Stanozolol
Mesterolone	
Methandriol	
Methenolone	
Methyltestosterone	
Nandrolone	
Norethandrolone	
Oxymesterone	
Oxymetholone	
Stanozolol	
Testosterone/epitestosterone	

[a] The anabolic steroid can be monitored as either parent or metabolite(s) in the listed fraction using the selected method.

3. Dry the ether extract by rotary evaporation or under nitrogen.

4. Derivatize with MSTFA:TMCS (100:2) at 60°C for 5 min.

5. Inject 1 μL into the GC/MS under the following conditions:

Gas Chromatograph: Hewlett-Packard 5890

   Column: 2-m HP Ultra 1 (SE-30, OV-1, OV-101), 0.2 mm ID, 0.33 μm film thickness

Injector temperature: 280°C

Injector system: Splitless

Initial temperature: 200°C (1 min hold)

Temperature rate: 30°C/min

Final temperature: 320°C (2 min hold)

Transfer line: 300°C

Carrier gas: He 1.0 mL/min

Mass Spectrometer: Hewlett-Packard 5970 or 5988

5.2   Conjugated Fraction Procedure

1. To 5 mL urine add 50 μL internal standard (methyltestosterone, 10 μg/mL).

2. Prewash C_{18} extraction columns with 5 mL methanol, then H_2O.
3. Aspirate urine through the columns and wash with 5 mL H_2O.
4. Elute with 5 mL methanol and collect eluate.
5. Dry the eluate by rotary evaporation or under nitrogen.
6. Add 1 mL 0.2 M sodium acetate buffer (pH 5.2) and vortex.
7. Hydrolyze the extract by incubating with 200 μL β-glucuronidase (*Helix pomatia*) at 60°C for 3 hr.
8. Cool to room temperature and add 100 mg Na_2CO_3/K_2CO_3 solid buffer (3:2) and vortex.
9. Add 1 g anhydrous Na_2SO_4 to each tube and extract with 5 mL diethyl ether.
10. Dry the extract by rotary evaporation or under nitrogen.
11. Derivatize with MSTFA:TMIS (1000:2 with 2 mg/mL dithioerythritol) at 60°C for 15 min.
12. Inject 1 μL into the GC/MS under the following conditions:

Gas Chromatograph: Hewlett-Packard 5890

> *Column:* 12-m HP Ultra 1 (SE-30, OV-1, OV-101), 0.2 mm ID, 0.33 μm film-thickness
> *Injector system:* Splitless
> *Injector temperature:* 280°C
> *Temperature 1:* 180°C
> *Rate 1:* 4°C/min
> *Temperature 2:* 228°C
> *Rate 2:* 2°C/min
> *Temperature 3:* 236°C
> *Rate 3:* 15°C/min
> *Temperature 4:* 280°C (2 min hold)
> *Rate 4:* 30°C/min
> *Temperature 5:* 320°C (3 min hold)
> *Transfer line:* 300°C
> *Carrier gas:* He 1.0 mL/min

Mass Spectrometer: Hewlett-Packard 5970 or 5988.

5.3. Confirmation

An important aspect of anabolic steroid screening is the confirmation of results. As with all forensic urine testing, it is a requirement to reanalyze the presumptive positive and compare the reanalysis with a known positive. In drug of abuse testing, this involves spiking a urine with either parent compound or in some cases metabolites, which are readily available. However, very few anabolic steroids are excreted without being metabolized and the metabolites and parent compounds for most anabolics are not available from commercial sources. Therefore, it is necessary to conduct metabolic studies, not only to determine the metabolic fate of these compounds, but to have a ready supply of "positive" urine for confirmation of screening results and assay quality control.

The following steps are minimal recommendations that should be performed for confirmational analysis.

1. Concommitant analysis of a blank urine, presumptive positive urine, and control urine should be performed with the matching of both spectral characteristics (full spectrums or appropriate ion ratios using at least four ions) and retention times of the anabolic steroid. The absence of interfering substances should be demonstrated by the analysis of the blank urine.

2. For the quantitation of the testosterone/epitestosterone (T/E) ratio, at least three aliquots of the presumptive positive urine should be assayed, along with a blank urine and a standard curve of testosterone/epitestosterone. The hydrolysis of the conjugated fraction should be done with β-glucuronidase derived from *E. coli*. The standard curve should span the cutoff level of 6:1. The mass spectra of testosterone and epitestosterone should be monitored to demonstrate noninterference from other compounds. Peak area and peak height determinations using the molecular ion, 432 *m/z*, should be performed and corrected by the standard curve.

5.4. Selected Chromatograms

Using the procedures described above, the free fraction assay of a urine containing methanedienone metabolites is shown in Fig. 10.6. Peak 1 is the 17-epimethandienone metabolite and Peak 2 is the 6-OH-methandienone metabolite. The stanozolol internal standard is present as the di-TMS (O-TMS, N-TMS) derivative. Figure 10.7 is the total ion chromatogram (TIC) of a normal blank conjugated fraction with the common endogenous compounds identified and listed in Table 10.4. This blank urine chromatogram can be compared to the TIC of a control urine in Fig. 10.8, which contains 11 anabolic steroids or metabolites. This control urine is analyzed with each group of urine specimens and is created from a mixture of metabolic urines and standards.

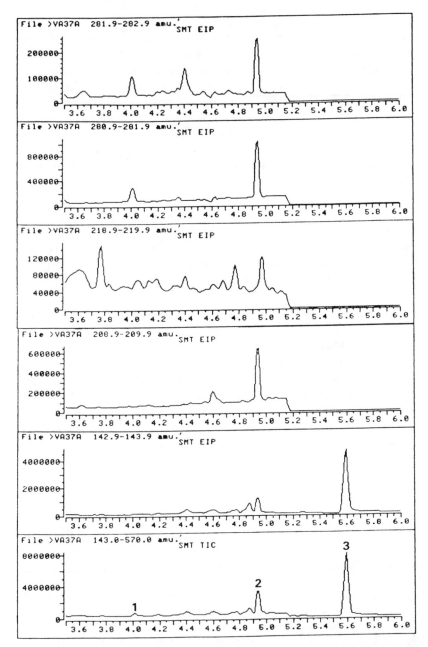

Figure 10.6. Extracted ion current profiles of free fraction extract. Selected ion monitoring chromatogram of urine containing methandienone metabolites. Peak 1: 17-epimethandienone TMS ether (4.0 min, 282.2, 219.2, and 143.2 *m/z*). Peak 2: 6-hydroxymethandienone bis(TMS ether) (4.9 min, 281.2, 209.2, and 143.2 *m/z*). Peak 3: Stanozolol *N*-TMS–*O*-TMS internal standard (5.6 min, 143.2 *m/z*).

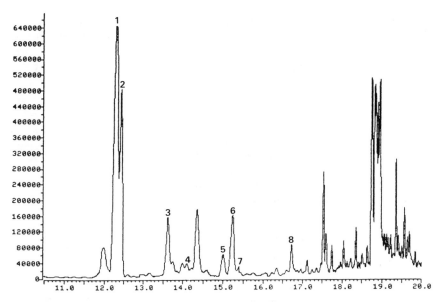

Figure 10.7. Total ion chromatogram of conjugated fraction extract of normal blank urine. Refer to Table 10.4 for identification of numbered peaks.

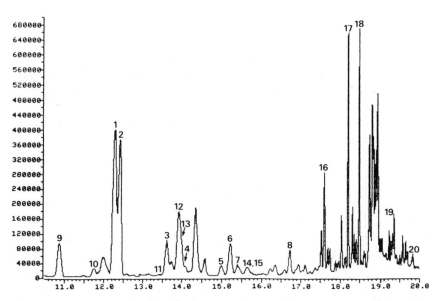

Figure 10.8. Total ion chromatogram of conjugated fraction extract of a conjugated steroid control. Pooled conjugated steroid control containing nine different anabolic steroid metabolic studies and spiked standards. Refer to Table 10.4 for identification of numbered peaks.

Table 10.4. Identification of Peaks in Figures 10.7–10.9

Peak Number	Compound
1	Androsterone
2	Etiocholanolone
3	Dehydroepiandrosterone
4	Epitestosterone
5	Testosterone
6	11β-OH-androsterone
7	11β-OH-etiocholanolone
8	Internal standard
9	Nandrolone metabolite 1
10	Nandrolone metabolite 2
11	Methenolone metabolite
12	Methyltesterone metabolite
13	Mesterolone metabolite
14	Chlorotestosterone
15	Norethandrolone/ethylestrenol metabolite 1
16	Norethandrolone/ethylestrenol metabolite 2
17	Oxymesterone
18	Oxymetholone metabolite
19	Stanozolol metabolite 1
20	Stanozolol metabolite 2

Figure 10.9 illustrates the computer-extracted data from the monitored ion at 432 m/z, which is used for the testosterone/epitestosterone ratio screen. A computer program is used to calculate the T/E ratio automatically and present area ratios relative to the internal standard that can be used for quantitative assessment. An elevated T/E ratio analysis is shown at the bottom of Fig. 10.9.

6. SUMMARY

The analysis of anabolic steroids in biological specimens has become an important part of doping control programs in athletics over the past 10 years. There is great concern over the misuse of these compounds for performance enhancement in athletes, and despite ongoing testing the use of these drugs by athletes persists. Education of athletes concerning the medical risks of these drugs is occurring at all levels of amateur sport, but additional research is needed on the long-term effects of these drugs. One of the steps being proposed by athletic organizations such as the

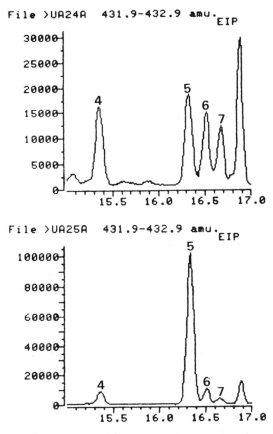

Figure 10.9. Extracted ion current profile for T/E ratio. Ion Chromatogram (432.2 *m/z*) used for calculation of T/E ratio. (Top) Normal T/E ratio (1:1). (Bottom) Elevated T/E ratio (12:1). Refer to Table 10.4 for identification of numbered peaks.

National Collegiate Athletic Association and the International Amateur Athletic Federation to combat the misuse of steroids further is the in-training testing of athletes. This step is considered important because athletes may use steroids during training periods and then discontinue the drugs' use soon enough to prevent detection during competition related testing.

The analytical technology that has developed for the detection and identification of anabolic steroids has grown over the past decade and parallels the development of improved technology in the field of chromatography and mass spectroscopy. Continued enhancements of the procedures being used in this area are expected and needed so that analytical aspects of doping control will be able to continue to provide an effective tool for drug deterrence programs in sports.

REFERENCES

1. H. A. Haupt and G. D. Rovere, *Am. J. Sports Med.*, *12*, 469 (1984).
2. D. R. Lamb, *Am. J. Sports Med.*, *12*, 31 (1984).
3. A. J. Ryan, *Fed. Proc.*, *40*, 2682 (1981).
4. R. H. Strauss, *Clinics Sports Med.*, *3*, 743 (1984).
5. J. A. Bell and T. C. Doege, *Physician Sportsmed.*, *15*, 99 (1987).
6. R. H. Strauss, J. E. Wright, G. A. M. Finerman, and D. H. Catlin, *Physician Sportsmed.*, *11*, 87 (1983).
7. M. Alen and P. Rahkila, *Int. J. Sports Med.*, *15*, 99 (1984).
8. B. F. Hurley et al., *J. Am. Med. Assoc.*, *252*, 507 (1984).
9. M. Moldawer, *J. Am. Med. Women's Assoc.*, *23*, 352 (1968).
10. R. H. Williams, Ed., *Textbook of Endocrinology*, 6th ed., Saunders, Philadelphia, 1981, p. 302.
11. M. Briggs and J. Brotherton, *Steroid Biochemistry and Pharmacology*, Academic Press, London, 1970, pp. 52–85.
12. R. J. Ward, C. H. L. Shackleton, and A. M. Lawson, *Br. J. Sports Med.*, *9*, 93 (1975).
13. R. Masse, C. Laliberte, L. Tremblay, and R. Dugal, *Biomed. Mass Spect.*, *12*, 115 (1985).
14. I. Bjorkhem and H. Ek, *J. Steroid Biochem.*, *18*, 481 (1983).
15. L. L. Engel, J. Alexander, and M. Wheeler, *J. Biol. Chem.*, *231*, 159 (1958).
16. D. F. Dimick, M. Heron, E. Baulieu, and M. Jayle, *Clin. Chim. Acta*, *6*, 63 (1961).
17. F. Galletti and R. Gardi, *Steroids*, *18*, 39 (1971).
18. E. Castegnaro and G. Sala, *Steroids Lipids Res.*, *4*, 184 (1973).
19. P. M. Adkikary and R. A. Harkness, *Acta Endocrinol.*, *67*, 721 (1971).
20. I. Bjorkhem, O. Lantto, and A. Lof, *J. Steroid Biochem.*, *13*, 169 (1980).
21. A. Karim, et al., *Clin. Pharm. Ther.*, *14*, 862 (1973).
22. H. W. Durbeck, I. Buker, B. Scheulen, and B. Telin, *J. Chromatogr.*, *167*, 117 (1978).
23. H. W. Durbeck and I. Buker, *Biomed. Mass Spect.*, *7*, 437 (1980).
24. E. L. Rongone and A. Segaloff, *J. Biol. Chem.*, *237*, 1066 (1962).
25. E. L. Rongone and A. Segaloff, *Steroids*, *1*, 179 (1963).
26. R. Rembiesa, et al., *Endocrinology*, *81*, 1278 (1967).
27. V. K. Schubert and K. Wehrberger, *Endokrinologie*, *55*, 257 (1970).

28. V. K. Schubert and G. Schumann, *Endokrinologie, 56,* 1 (1970).

29. R. V. Quincey and C. H. Gray, *J. Endocrin., 37,* 37 (1967).

30. C. J. W. Brooks, A. R. Thawley, P. Rocher, B. S. Middleditch, G. M. Anthony, and W. G. Stillwell, *J. Chromatogr. Sci., 9,* 35 (1971).

31. R. Dugal, C. Dupruis, and M. J. Bertrand, *Br. J. Sports Med., 11,* 162 (1977).

32. W. R. Jondorf, *Experientia, 36,* 394 (1980).

33. R. V. Brooks, R. G. Firth, and N. A. Sumner, *Br. J. Sports Med., 9,* 89 (1975).

34. R. Hampl and L. Starka, *J. Steroid Biochem., 11,* 933 (1979).

35. R. V. Brooks, G. Jeremiah, W. A. Webb, and M. Wheeler, *J. Steroid Biochem., 11,* 913 (1979).

36. N. A. Sumner, *J. Steroid Biochem., 5,* 307 (1974).

37. V. Mann, A. B. Benko, and L. T. Kocsar, *Steroids, 37,* 593 (1981).

38. L. Laitem, P. Gaspar, and I. Bello, *J. Chromatogr., 147,* 538 (1978).

39. K. L. Oehrle, K. Vogt, and B. Hoffman, *J. Chromatogr., 114,* 244 (1975).

40. R. Verbeke, *J. Chromatogr., 177,* 69 (1979).

41. A. T. Rhys Williams, S. A. Winfield, and R. C. Belloli, *J. Chromatogr., 240,* 224 (1982).

42. E. H. J. M. Jansen, R. Both-Miedema, H. Van Blitterswijk, and R. W. Stephany, *J. Chromatogr., 299,* 450 (1984).

43. E. H. J. M. Jansen, P. W. Zoontjes, H. Van Blitterswijk, R. Both-Miedema, and R. W. Stephany, *J. Chromatogr., 319,* 436 (1985).

44. E. H. J. M. Jansen, H. Van Blitterwijk, P. W. Zoontjes, R. Both-Miedema, and R. W. Stephany, *J. Chromatogr., 347,* 375 (1985).

45. C. G. B. Frischkorn and H. E. Frischkorn, *J. Chromatogr., 151,* 331 (1978).

46. R. H. Van Den Berg, E. H. J. M. Jansen, G. Zomer, C. Enkelaar-Willemsen, R. Both-Miedema, and R. W. Stephany, *J. Chromatogr., 360,* 449 (1986).

47. S. Siggia and R. A. Dishman, *Anal. Chem., 42,* 1223 (1970).

48. R. A. Henry, J. A. Schmit, and J. F. Dieckman, *J. Chromatogr. Sci., 9,* 513 (1971).

49. F. A. Fitzpatrick, *Clin. Chem., 19,* 1293 (1973).

50. G. Carignan and B. A. Lodge, *J. Chromatogr., 179,* 184 (1979).

51. E. C. Horning, M. G. Horning, N. Ikekawa, E. M. Chambaz, P. I. Jaakonmaki, and C. J. W. Brooks, *J. Gas Chromatogr., 283,* (1967).

52. K. D. R. Setchell, B. Alme, M. Axelson, and J. Sjovall, *J. Steroid Biochem., 7,* 615 (1976).

53. M. Axelson, B. L. Sahlberg, and J. Sjovall, *J. Chromatogr., 224,* 355 (1981).

54. J. Sjovall and M. Axelson, *Vitamins Hormones, 39,* 31 (1982).

55. H. L. Bradlow, *Steroids, 11,* 265 (1968).

56. H. L. Bradlow, *Steroids, 30,* 581 (1977).

57. J. Sjovall and M. Axelson, *J. Steroid Biochem., 11*, 129 (1979).

58. J. Sjovall, *J. Steroid Biochem., 6*, 227 (1975).

59. B. E. P. Murphy and R. C. D. D'Aux, *J. Steroid Biochem., 6*, 233 (1975).

60. C. H. L. Shackelton and J. O. Whitney, *Clin. Chim. Acta, 107*, 231 (1980).

61. J. P. Thenot and E. C. Horning, *Anal. Lett., 5*, 21 (1972).

62. J. P. Thenot and E. C. Horning, *Anal. Lett., 5*, 801 (1972).

63. L. L. Engel, A. M. Neville, J. C. Orr, and P. R. Raggatt, *Steroids, 16*, 377 (1970).

64. M. Axelson and J. Sjovall, *J. Steroid Biochem., 5*, 733 (1974).

65. M. Bertrand, R. Masse, and R. Dugal, *Farm. Tijdschr. Belg., 55*, 85 (1978).

66. E. M. Chambaz, C. Madani, and A. Ros, *J. Steroid Biochem., 3*, 741 (1972).

67. L. Aringer, P. Eneroth, and J. A. Gustafsson, *Steroids, 17*, 377 (1971).

68. M. Donike and J. Zimmerman, *J. Chromatogr., 202*, 483 (1980).

69. V. P. Uralets, V. A. Semenova, M. A. Yakushin, and V. A. Semenov, *J. Chromatogr., 279*, 695 (1983).

70. T. Feher and E. H. Sarfy, *J. Sports Med., 16*, 165 (1976).

71. I. Bjorkhem and H. Ek, *J. Steroid Biochem., 17*, 447 (1982).

72. O. Lantto, I. Bjorkhem, H. Ek, and D. Johnston, *J. Steroid Biochem., 14*, 721 (1981).

73. M. Donike, J. Zimmermann, K. R. Barwald, et al., *Deutsch Z. Sportmedizin, 35*, 14 (1984).

74. G. P. Cartoni, M. Ciardi, A. Giarrusso, and F. Rosati, *J. Chromatogr., 279*, 515 (1983).

75. D. H. Catlin, R. C. Kammerer, C. K. Hatton, M. H. Sekera, and J. L. Merdink, *Clin. Chem., 33*, 319 (1987).

76. C. K. Hatton and D. H. Catlin, *Clinics Lab. Med., 7*, 655 (1987).

77. M. Donike, *Proceedings of 4th Cologne Workshop on Dope Analysis*, Cologne, West Germany, April 6–11, 1986.

CHAPTER

11

DRUG ANALYSIS TECHNOLOGY—PITFALLS AND PROBLEMS OF DRUG TESTING*

ARTHUR J. McBAY

*University of North Carolina,
Chapel Hill, North Carolina*

1. INTRODUCTION

Testing urine for drugs of abuse has become both popular and controversial (1, 2). The amount of money available has made such testing very

---

* A similar version of this chapter under the same title was originally published by the American Association for Clinical Chemistry in the proceedings of the 1987 Arnold O. Bockman Conference in Clinical Chemistry entitled "Drug Abuse in the Workplace: Prevention and Control," *Clin. Chem., 33*(11B), 33B–40B (1987).

profitable. The demand for such large numbers of tests on urines obtained randomly from symptomfree subjects requires many technicians, many of whom are not familiar with the care needed to process such specimens or with the procedures required to protect the rights of those involved.

Tests are available to identify properly and to quantify the analytes that advocates of such testing say should be determined. The tests can be divided into two types of procedures: (1) the presumptive screening tests, which are intended to differentiate easily, rapidly, and inexpensively specimens that most probably contain the sought-for analytes from those that do not, and (2) the tests that can confirm the identification. The latter tests may be complex, slow, and expensive but they are essential because they can provide adequate documentation if necessary for adversarial proceedings. Immunoassays and chromatographies are considered as presumptive screening tests, and gas chromatography/mass spectrometry (GC/MS) can be a confirmatory procedure. Of course both kinds of tests must be performed properly to fulfill these roles.

The sudden increase in demand for urine drug testing is based largely upon anecdotal information, or in many instances, upon attributing blame for incidents to a drug, based on the presence of the drug in the person involved or on surveys of reported drug use. Military testing expanded exponentially after the finding of an antihistamine in a pilot who crashed during a night landing on a carrier, killing several seamen whose urines tested positive for cannabinoids. A fatal train crash was related to marijuana because cannabinoids were found in the operating crewmen, who reportedly had fallen asleep while operating the train (3). In studies of fatally injured motor vehicle operators, alcohol was found in 64% of the drivers, drugs with alcohol in 28%, and drugs alone in 7%. No evidence was found that indicated that a significant number of drivers might have been impaired by drugs other than alcohol (3). A survey reported that the admitted incidence of drug abuse in the military decreased in 1985 over that reported in 1980. However, testing procedures prior to 1984 were questionable (4). The survey also reported that alcohol use increased. Another survey on the economic cost to society of drug abuse has been cited as evidence of the problem of drug abuse. The survey makes an excellent case for the very large economic cost of alcohol abuse but makes no case for testing for cannabinoids and other drugs in the workplace except for alcohol (5).

Industries that have testing programs and advocates who sell such programs are quoted in the media to the effect that significant improvements in the drug problem have followed the introduction of these programs. However, no data or documentation is offered describing the extent of the problem before or after drug testing or indicating that programs are cost-effective.

2. IMMUNOASSAYS

The use of immunoassays, the tests most often used and contested in adversarial matters involving drug testing in urine, is limited in scope, because these tests are designed to detect only one analyte or a few closely related ones. Separate tests must be performed for each analyte but, because the same procedure is used for all of the analytes the tests have been designed to assay, one need only change reagents to test for different analytes. Technician training is minimal and the tests can be automated. However, the inexperienced technician may not know when the assay is out of control or how to interpret the findings.

The reliability of an immunoassay depends upon the specificity of the antibody, the means of detecting the drug and (or) its metabolites, the variability of the detection signal, and the capability of the personnel doing the testing.

Immunoassays depend on the binding of an analyte to an antibody. The specificity of the prepared antibody toward an analyte depends on how the analyte was attached to the antigen carrier. By changing the point of attachment of the antigen on the analyte, one can change the cross-reactivity of an antibody. When immunoassays are produced with different lots of antibodies, the cross-reactivities should be redetermined. Specificity also depends on whether the analyte has a unique structure that differs significantly from any other structure that may be present in the specimen. EMIT®-Opiates and Abuscreen®-Morphine assays significantly cross-react with codeine. Abuscreen and EMIT amphetamine assays cross-react with several phenethylamine drugs. EMIT and Abuscreen Cannabinoid assays, designed to detect 11-nor-Δ^9-tetrahydrocannabinol-9-carboxylic acid (THC-COOH), also cross-react with other cannabinoids.

2.1. Homogeneous Enzyme Immunoassays

The EMIT assay is a homogeneous assay. That is, it has the advantage of not requiring a physical separation step. An enzyme to which analyte molecules have been permanently attached is used as a marker. The antibody binds the enzyme–analyte combination as if it were an analyte. When a drug is present in urine, these enzyme markers are proportionately displaced into the solution, where their reaction with an appropriately chosen reagent changes the absorbance of light passing through the solution. Anything that affects the enzyme activity, such as temperature changes and substances other than the analyte being sought (e.g., cannabinoids), can produce erroneous results. The threshold (cutoff) that separates positive from negative test results is predetermined for immunoassays by measuring the amplitude of the signal obtained from a known amount of the drug. If the cutoff is set too low, that is, if the amplitude is too close to that of a negative specimen, then a confirmatory test (e.g., GC/MS, which specifically tests for one compound of those that might respond to an immunoassay) may not confirm the result of the first test. If the cutoff is set too high, some positive specimens may be missed but any confirmations should be more reliable. With adequate quality control, cutoff values can be established for laboratory procedures or for specific analysts (6).

The terms "false positive" and "false negative" are used to characterize the reliability of the assay. A false positive means that a positive response to the assay is produced when the analyte being sought is not present, is not confirmable as present, or is present at a concentration below that of the cutoff. This may occur also if some other substance is identified as the analyte, for example, if the specimen has been contaminated or mislabeled or if pipettes were not thoroughly cleaned between tests, particularly when the preceding sample tested contained a large concentration of the analyte. A false negative means that the drug is present in a concentration greater than the cutoff concentration, yet gives results less than the cutoff concentration. False negatives may result from an interference, the loss of analyte, or technician error such as a pipetting error. If a test does not confirm the presence of a drug, there is little interest in expending time and money to try to find whether that is, in fact, the case and what is the cause of the false negative. Information concerning false positives also is not readily available. Tests that are not confirmed are not going to reveal false positives.

2.1. Radioimmunoassays

Radioimmunoassays use markers that are radio-labeled derivatives of the drug or metabolite. The analyte in urine will proportionately displace the radiolabeled drug. The displaced and the undisplaced drugs must be physically separated. The amount of radioactivity of the displaced or the undisplaced drug can be determined and compared with controls similarly treated to determine the amount of drug in urine. If the separation is poor, the results will be incorrect.

Both the enzyme immunoassay and the radioimmunoassay results and interpretation depend on the amplitude of a signal for determining whether a sample is positive or negative for a drug. This in turn depends on the efficiency of measurements and techniques. Using properly controlled automated instruments can minimize errors of technique and measurement. The quality of results from nonautomated procedures depends more on the expertise of technicians and the quality control of the analyses.

2.3. Interpretation of Cutoff Concentrations

Immunoassays for cannabinoids are usually standardized against THC-COOH, which represents about 25% (range 0–60%) of the cannabinoids in urine, but they also respond to other cannabinoids present. A specimen that gives a positive response equivalent to that for 100 μg/L of THC-COOH in an immunoassay may contain 0–60 μg/L of total THC-COOH. A 20 μg/L immunoassay-positive urine may contain 0–12 μg/L of total THC-COOH. Indeed I recommend discouraging the use of the 20 μg/L cutoff immunoassay because of the decreased certainty of the drug-positive identification and the increased number of tests with unconfirmable results that are produced.

The chromatographic methods presumably respond to THC-COOH. The amount of this analyte detected in the specimen depends on whether the procedure detects both free and conjugated THC-COOH and on the possibility of interference from the other metabolites and other substances present. The GC/MS method, which is measuring only the THC-COOH, usually gives a lower value than the estimate made with an immunoassay. It would be rare for a semiquantitative estimate based on an immunoassay to give a concentration similar to that obtained by GC/MS. The performance characteristics of immunoassays are based on comparison with GC/MS results or by testing drugfree urines and comparing them with the same urine with known amounts of analyte added. In tests of the EMIT d.a.u. Cannabinoid 20 Assay with 25–50 cannabinoid-free urines and 25–50 urine samples with 50 μg/L THC-COOH added (7), the assay correctly identified more than 99% of these cannabinoid-supplemented specimens. Had the assay been tested against specimens with only 20 μg/L THC-COOH added, we could expect from gaussian distribution that the assay would have correctly identified 50% of the specimens. Decisions on choosing cutoffs depend on how many false negatives, false positives, and unconfirmed tests are economically and scientifically acceptable.

2.4. Confirmation of Results

The identification of the measured apparent compound in the specimen must be confirmed to be the same as the known analyte in the reference standard. Identification and quantification should not be stated as confirmed if the calculated concentration is less than that of the lowest standard sample. Analyte added to a blank sample may behave differently from analyte naturally present in a specimen. In a recovery study a higher recovery of the analyte in the control than in the specimen would lead to erroneously low results. Also, analytes present in a specimen can escape detection or can produce false positives. If the analyte concentration is below the sensitivity of the assay, the result should be reported as negative or unconfirmed. A subject may take large amounts of fluids, which can dilute the concentration of analyte in urine below that of the cutoff. The addition of salt, vinegar, hypochlorite bleach, liquid detergent, liquid soap, or blood to known cannabinoid-positive urine specimens has produced false-negative EMIT results (8). Without further testing, the subjects' urines would be considered negative for the tested analytes.

2.5. Passive or Involuntary Ingestion of Drugs of Abuse

Cannabinoids have been found in urine and blood after the passive inhalation of marijuana smoke (9, 10), and the positive urine immunoassay results have been confirmed. The concentrations found depend on the amount of the exposure. Cannabinoids have also been found in urine and blood after the oral ingestion of foods containing cannabis resin or hashish (11). Subjects may be unaware of the intake of marijuana and feel no effects but their urines may contain cannabinoids for several days.

2.6. Interferents

Positive cannabinoid results by EMIT have been obtained from urines of people who have been taking the anti-inflammatory drugs ibuprofen (Motrin, Advil, etc.), naproxen, and fenoprofen, which therefore may potentially interfere with EMIT assays for amphetamine, barbiturate, benzodiazepine, and methaqualone (12). The manufacturer of the tests has since changed the assays and claims that this is no longer a problem. The importance of these false positive findings is that millions of assays have been performed in the more than seven years that ibuprofen, a most popular and widely used drug, has been available. Any of the many EMIT assay results that were unconfirmed could have been reported as positive, even when marijuana had not been used. It is unlikely that a scientist could forecast that such relatively simple phenylpropionic acid derivatives would produce false positives in urine screening tests for cannabinoids and other drugs. This interaction should not be a problem when adequate confirmatory tests are used, but many immunoassays are still not being confirmed.

Although many substances could interfere with any assay to produce erroneous results, little is known about what substances do interfere. There has been practically no interest in doing further testing on specimens that screened negative or that gave positive results in screening tests but were not confirmable.

The morphine contained in poppy seeds can be detected in urine from subjects who eat foods containing the seeds (13). The use of codeine pharmaceuticals can result in the finding of morphine in urine, and if the test is sensitive enough, codeine might be found also. Because these results are not false positives, one would expect that they would be confirmed in followup testing.

Benzoylecgonine, a metabolite of cocaine, was measured at 1200 μg/ L in urine from a subject who had drunk one cup of "Herbal Inca Tea" (14). This product was alleged to contain decocainized coca leaves but in fact contained about 5 mg of cocaine per tea bag (15).

The antibodies used in amphetamine immunoassays can cross-react with other phenethylamine compounds such as methamphetamine, ephedrine, phenylpropanolamine, and pseudoephedrine. Although the amphetamines are controlled substances, three of the other substances are sold in widely used over-the-counter preparations. Most users of these preparations are unaware of what they contain or that they may cause false positive urine tests. Adequate confirmation should prevent the problem of misidentification.

Barbiturate immunoassays do not differentiate among the various barbiturates that are available and widely used. This can create problems because finding phenobarbital in urine could indicate use of a drug that was (a) prescribed as a mild sedative, (b) used for treating epilepsy, or (c) obtained in a nonprescription anti-asthmatic preparation in combination with ephedrine. Ephedrine also can produce a positive response to an amphetamine assay. Butalbital is present in analgesic preparations, sometimes combined with codeine. Amobarbital, pentobarbital, and secobarbital are used as sedative-hypnotics. Confirmation methods should identify the specific barbiturate. These drugs could impair performance, but measuring their urine concentrations would not establish this. Measuring blood concentrations might.

3. SPECIFICITY OF CHROMATOGRAPHIC ASSAYS

Design of immunoassays for a single drug or for closely related compounds lends a certain amount of specificity to these assays. Chromatographic methods are useful if a large number of drugs are sought, but an identical response may be produced by many compounds.

3.1.   Thin-Layer Chromatography

Thin-layer chromatography (TLC) is useful for screening urines because multiple specimens can be applied to a chromatographic plate and more than one drug can be determined for each application. The disadvantages are that the procedure may require a prior separation, purification, or concentration of the specimen and technical skill for careful application of samples to the plate. After the solvent front reaches a certain point on the plate, drugs are usually made visible by spraying the dried plate with various reagents. The positions and appearances of any spots are used to identify the drug or drugs being sought. The certainty of the identification depends on the efficiency of the procedure, the ability of the technicians performing the assays, and the ability of those making the identification. Color photographs or the actual chromatograms are the only evidence that can serve as documentation of the quality of the work. The procedure can be an excellent and relatively inexpensive screening test, sometimes for multiple drugs, but it can lack the specificity of immunoassays. Most urines contain substances that can produce spots on thin-layer chromatograms so as to mask or be misinterpreted as the analyte being sought. Adequate confirmation should prevent the reporting of false positives.

3.2.   Gas Chromatography and High Performance Liquid Chromatography

Gas chromatography (GC) and high performance liquid chromatography (HPLC) are also useful for screening. Extracts of urines are injected into these instruments onto columns through which gases or liquids are flowing. The columns separate the sample constituents, and the various times that each constituent takes to affect the detector and the amplitudes (peaks) of the signal are recorded. The identification is based on this time to peak elution, called the retention time. This system should ideally distinguish THC-COOH from about 50 other compounds chosen at random. Assuming a 6-min run and allowing a retention time variation of ± 0.06 min, the great majority of the chromatographically separable compounds will fall within 50 windows (6.0/0.12) of run time (16).

With some methods the drug produces two responses and its metabolite two more. The increased number of responses will increase the possibility of accurate identification. An example of this would be the use in HPLC of the relative peak heights of cocaine and benzoylecgonine determined at two different ultraviolet wavelengths for each analyte. Greater specificity is also possible with HPLC by performing several immunoassays on fractions obtained before, during, and after the suspected analyte is expected to peak (17). The use of additional independent testing methods can decrease the number of false positives. Because the documentation in gas and liquid chromatography consists of retention times and peak heights that are not unique, chromatographic methods are not generally accepted as adequate for confirming identification in adversarial proceedings.

3.3. Gas Chromatography/Mass Spectrometry

Mass spectrometry (MS) is used in the procedure generally accepted by the scientific community for the confirmed identification of analytes. Other methods may be acceptable in some cases or at some time in the future. In an adversarial proceeding where an individual's freedom, reputation, and career are in jeopardy, the specific information and documentation possible by using properly performed GC/MS are essential. Recently, mass spectrometry coupled with liquid chromatography or with a second mass spectrometer has led to methods that will be adequate enough for identification. The GC/MS methods can provide a record of the complete mass spectrum of an analyte that appears at a particular retention time. One can usually distinguish if more than one analyte appears at the same retention time. The mass spectrometer can be programmed to record several mass ions or a single mass ion. The fewer the mass ions sought, the greater the sensitivity, but the less the probability of accurate identification. The total mass spectrum, if enough mass ions of the proper intensities are included, should constitute enough evidence for an identification without any other evidence. In most instances, the specimen would have given a positive screening result beforehand. The combination of a positive immunoassay and a GC/MS result with the proper retention times and three mass ions of the proper intensities should distinguish the compound of interest from more than a million other organic compounds. Two mass ions of the proper intensities will distinguish the compound from about 10,000 compounds. A single mass ion would distinguish the compound from a smaller number of compounds (16). The likelihood of identification may be increased by extracting, separating, derivatizing, using different ion sources, increasing resolution, and using improved packed or capillary columns.

4. QUALITY ASSURANCE

4.1. Problems of Misidentification

An example of the problem of a possible misidentification has been reported (18). Cyclobenzaprine, a skeletal muscle relaxant, differs structurally from amitriptyline, a tricyclic antidepressant, by the addition of a double bond to the central ring of amitriptyline. Nevertheless, both compounds give identical responses to three different TLC procedures, have the same retention times with GC and HPLC procedures, give positive EMIT responses, and produce a GC/MS ion of m/e 275. A GC/MS method of identification (19) is available that produces some different mass ions for the two compounds.

One spot test kit for screening urine specimens for drugs has been advertised, the KDI Quik Test-Narcotic Screen (20). Urine is drawn into a syringe containing material that absorbs and separates some dissolved substances. A few drops of the eluate are placed onto a circle of iodo-platinate-impregnated test paper. The appearance of a blue-gray ring after the test paper is washed supposedly indicates a positive test for "abused drugs" (21). The test is marketed as being useful for screening out negative urine specimens, but the stated sensitivity of 2000 μg/L for phencyclidine (PCP), 3000 μg/L for amphetamines and morphine, and 5000 μg/L for "cocaine" might produce many false negatives. The most common screening methods, immunoassays, are 3 to more than 25 times more sensitive. In addition to the above drugs, the test produces the same positive response to unstated concentrations of benzodiazepines, phenothiazines, amitriptyline, propoxyphene, methadone, flurazepam, quinine, oxymetazoline, phenylpropanolamine, clorazepate, codeine, pyrilamine, dextromethorphan, prochlorperazine, methamphetamine, and other substances, but not to cannabinoids and alcohol. It is uncertain how many other drugs have been tested, but many organic bases appear to give the identical positive responses. Again the positive responses should be screened by another method to establish what method is needed to identify and confirm the drug or drugs present. This method is nonspecific, whereas TLC methods partially separate analytes or can produce color changes when plates are sequentially sprayed with various reagents. Trying to confirm which of the potentially great number of positives is present could make this a very expensive test. Positive results that are not confirmable could be damaging to the person tested.

According to a recently proposed theory, blacks have more melanin in their urines than whites, and this melanin may be misidentified as THC-COOH by GC/MS procedures (22). Consequently, a "de-melanizing solution'' has been prepared and tested. The urine of a black serviceman calculated to contain about 5 μg/L of THC-COOH was de-melanized and retested; however, about 90% of the 20 μg/L internal standard (deuterated THC-COOH) apparently also disappeared along with the evidence of the THC-COOH. Examination of the urines of three more black servicemen revealed quantitatively insignificant amounts of the mass ions used in marijuana determinations. Thus the available information does not support the theory that urines from blacks yield false positives or increases in the calculated amounts of THC-COOH because of melanin. However, this work does support the allegations of difficulties that occur when specimens that supposedly contain concentrations of THC-COOH less than 20 μg/L are analyzed routinely.

4.2. Court Evidence

The admissibility of urine drug testing results in courts may be determined by the Frye standard (23) by the Federal Rules of Evidence (24), or both. Under the Frye standard, the scientific procedure being considered for admission in evidence must be sufficiently established to have gained general acceptance in the particular field in which it belongs. The Federal Rules of Evidence consider both the relevance and probativeness of a particular technique. Applying these rules for evaluating the procedures used in testing for drugs in urine, one could exclude immunoassays, TLC, GC, and HPLC, but MS methods should be acceptable (25).

4.3. Quality Control

Quality control of the confirmatory method is most important because in adversarial proceedings documentation of confirmations should have the most weight. Quality control of the screening procedure is important in maintaining an economical balance between false positives and false negatives. Laboratory personnel should not state that their performance is equal to that claimed by the manufacturers of reagents or instruments unless they have independently established this by adequate quality control. The limit of detection and the limit of quantification and qualitative confirmation should be established in a laboratory by each technician performing the test (6). The limit of detection is the lowest concentration of an analyte that the analytical process can reliably detect. The limit of quantification should be clearly above the limit of detection, and the numerical significance of the apparent analyte concentration should increase as the analyte signal increases above the limit of detection.

4.4. Proficiency Testing

The reliability of results of urine tests performed in the large number of laboratories offering such analyses has not been determined and is consequently unavailable at present. Results of tests of 13 laboratories in 1981 essentially agreed with the findings of three earlier studies (26). In all the studies the numbers of false positives and false negatives increased greatly when the specimens were submitted "blind" to a laboratory along with specimens from one of the laboratory's usual sources, as compared with the same specimens being submitted such that they could be identified as proficiency test specimens. The average correct response rate for six analytes on blind positive specimens was 46.5% (range 0–100%) versus 95% (68–100%) for known proficiency test specimens. The correct response rate was 96.8% (34–100%) on blind negative specimens and 99.3% (92–100%) on known specimens. One conclusion drawn was that the laboratory personnel were not applying their full capability to the analysis of routine specimens.

In testing urines the adequacy of the proper identification of analytes is of prime importance; quantification is important only if there is the possibility of establishing urine drug concentrations that could in any way be correlated with the amount of drug used, time of use, and alteration of performance, or if "per se" concentrations are requested. Consistency in quantitative control from specimen to specimen, and from day to day, would help provide accurate results for all those tested.

The results obtained from proficiency testing programs of the armed services laboratories have been available for review by both sides in adversarial proceedings. The program is mandatory for these laboratories and for their contract laboratories. Several states have proficiency testing programs for certain laboratories in their jurisdiction. The College of American Pathologists has voluntary proficiency testing programs for which fees are charged. The American Association of Bioanalysts is reported to have a urine toxicology program (27). There have been reports of private enterprise proficiency testing, but unfortunately the results have not generally been available to independent experts in adversarial proceedings. When private laboratory personnel have been asked about the accuracy of their tests, they have responded that their work is 95% or 99% accurate. Rather than offering documentation to support their claim, they cited a manufacturer's pamphlet describing testing performed on drug-supplemented specimens or some scientific paper published on testing similar to their method, but they offered no evidence of the quality of their own work.

Programs have been available for testing the proficiency of those analyzing blood for alcohol, but results obtained from such programs cannot be compared with those testing for other drugs. Blood is tested for a specific volatile substance, ethanol, that is present in relatively large concentrations. Most methods will distinguish ethanol from the other infrequently occurring interferences, methanol and isopropanol. An alcohol concentration of 0.10%, or 100 mg/dL, equals 1,000,000 μg/L. Urine concentrations of ethanol are about 1.3 times higher than those of blood. An excellent review of recent developments in alcohol analysis is available (28).

Some of the other drugs of abuse that might be found in urine are also usually present at much greater concentrations than, for example, cannabinoids. Barbiturates, opiates, cocaine (benzoylecgonine), amphetamines, and methadone are present at concentrations of about 1000 μg/L, whereas cannabinoids and PCP are about one-tenth this concentration (100 μg/L). In addition to being more concentrated in urine, the other analytes are present less frequently, and large numbers of similar analytes can cross-react in their immunoassays.

Obtaining drugfree urine to use as a matrix to which known amounts of drugs are added to provide proficiency testing specimens is not easy. When pooled urines are obtained from the general population, many drugs may be present but, it is hoped, in too low a concentration to be detected by some methods. Unfortunately, not only may a given drug be present and detectable but also adding a known amount of the same drug to the pooled urine may give a final concentration greater than expected. "Synthetic" urines pose problems in that the initial absorbance, which is not within the normal range for urines, may be rejected in an enzyme immunoassay and these matrices may lack electrolytes required for other immunoassays.

Stability of specimens and the reliability of "reference laboratories" in determining low concentrations of drugs also make the evaluation of participating laboratories difficult.

4.5. Methods Comparisons

Most methods comparisons are concerned with determining cannabinoids in urine. An exception is a study of proficiency testing results (29). The reported results for the identification and quantification of the various analytes were poor when judged by the numbers of false positives, false negatives, and the large variances in the quantitative results.

The proficiency of some armed services laboratories in testing a large number of specimens for cannabinoids has been reported by Abercrombie and Jewell (30), who found 91% agreement between the 100-μg/L cutoff

EMIT and the 100-μg/L cutoff Abuscreen RIA in the laboratory. Their data indicated a 4% false positive rate and a 10% false negative rate in EMIT field testing.

Six cannabinoid assays have been compared (31). Three of the four immunoassays used different antibodies with different cross-reactivities and cutoffs of 100, 20, and 6 μg/L for THC-COOH equivalents. The TOXI-LAB assay did respond positively to two urines containing 10 μg/L of THC-COOH added. The cutoff for the GC/MS method was 4 μg/L. The differences in cross-reactivities and cutoffs emphasize difficulties with trying to compare methods. For example, 2–3% of the positive EMIT-d.a.u. 20 μg/L results could not be confirmed.

A comparison between TOXI-LAB, EMIT-d.a.u., and EMIT-st concluded that both EMIT-d.a.u. and TOXI-LAB assured a reliable assay of urinary cannabinoids when the cannabinoid concentration exceeded 25 μg/L (32). However, only 5 of the 523 specimens were checked by GC/MS. In another study (33) 30 urines were found positive for cannabinoids by EMIT-d.a.u., TLC, and GC/MS procedures; similarly, 30 "true negatives" were negative by EMIT-d.a.u. and TLC procedures, but the GC/MS procedure detected spurious low concentrations of THC-COOH (approximately 5 μg/L) in two instances. Irving et al. (34) compared EMIT-st, EMIT-d.a.u., Abuscreen RIA, Research Triangle Park RIA, GC, and GC/MS methods. The GC method confirmed 69–92% of selected specimens found positive by any of the three commercial immunoassays, depending on the method and the cutoff used. The GC/MS method confirmed 98% of the EMIT and RIA positives with a 20-μg/L cutoff. Some of the earlier comparisons involved GC/MS procedures that did not determine conjugated THC-COOH, whereas the immunoassays detected both free and conjugated cannabinoids. This problem, plus the loss of analyte that can occur during extractions and the fact that the GC/MS method is determining only one of the cannabinoids (THC-COOH), led to low concentrations or negative findings by GC/MS. The value of hydrolysis and of using combined HPLC and RIA methods has been reported (17, 35).

5. TEST MENUS

Most of the screening tests have been for urinary cannabinoids, but recently interest in testing for a cocaine metabolite, benzoylecgonine, has increased. Immunoassays are available for amphetamines, barbiturates, benzodiazepines, methadone, methaqualone, PCP, and propoxyphene, but not much is heard about testing urine for these drugs or for confirming positive findings of such tests.

Some professional athletes are being tested for the presence of drugs in their urines but it is unclear as to how many, how often, or for what substances, or as to what effect testing has had on sporting events. Limited testing has been mandated by the National Collegiate Athletic Association for more than 80 substances (36). The costs of performing such tests once or twice for about 3000 athletes is about $1 million after a start-up cost of about $2 million (37). Test results are to be confirmed by GC/MS procedures.

Confirmatory methods for many of the above drugs are expensive, not readily available, and performed in very few laboratories. The logical extension of such testing would include drugs with a greater potential to impair performance such as alcohol, antidepressants, hydromorphone, fentenyls, methylenedioxymethamphetamine, buprenorphine, etorphine, lysergic acid diethylamide (LSD), and "designer" drugs. It will be difficult or impossible to find laboratories capable of routinely performing such testing, and this testing will be very expensive.

6. OTHER SPECIMENS

Serum and blood are good specimens for drug screening because active substances persist in the circulating blood for a shorter time after use and in some instances the concentrations of these substances might be correlated with their effect. Because these specimens are being taken directly from the patient by qualified personnel, substitution or contamination of the specimen by the patient is generally not possible. Analyses of the specimens are more difficult, more expensive, and more complex; the amount of specimen is smaller; and the concentrations of drugs and metabolites are lower.

Saliva has been proposed as a specimen. Many drugs are present in saliva after first being absorbed into the bloodstream (38), although THC is found in saliva owing to direct contact with marijuana smoke (39). Cocaine has been found in saliva in about the same concentrations as in plasma after intravenous injection (40). The disadvantages of using saliva are the difficulty of obtaining the specimen, small sample size, variability in duration of positives (41), and the difficulty or impossibility of confirmation and quantification.

Hair has been proposed as a specimen (42). Obtaining samples would presumably be less intrusive than obtaining urine and, by making several determinations along the hair, one could pinpoint the time of drug use. The disadvantages are similar to those of saliva: the difficulty or impossibility of confirmation and quantification. It is unlikely that recent drug use, within 24–48 hr, could be detected in hair specimens.

Perhaps urine should be obtained in addition to hair and saliva so that confirmation and quantification would be possible.

7.   INTERPRETATION OF RESULTS

Measurements of drugs in urine cannot be used to establish that performance or health is adversely affected; the presence of a drug in urine means only that the person must have ingested the drug. Marijuana illustrates the complexity of the problem. Marijuana cigarettes vary greatly in cannabinoid composition, even when the tetrahydrocannabinol (THC) concentration is known. People smoke different quantities at different rates, and some use the drug orally. Individuals absorb, distribute, metabolize, conjugate, store, and excrete the more than 30 metabolites of marijuana differently.

Nothing is more important than the interpretation of the results of drug urine testing, when they are to be used in adversarial matters. The objectives of the testing, which should be foremost in the minds of those doing the interpretation, are to assure health, safety, and performance in the workplace. An expert witness can give an opinion based on his or her own studies, or on the documentation of the work of others, that the specimen did not contain the substance sought at or above the sensitivity of the method of analysis.

When a drug or metabolite is reported as being present in urine, this means that the person was exposed to the drug voluntarily or involuntarily or that there was some error in the identification of the specimen or in the analytical procedure. The expert's opinion can dwell on the adequacy of the procedure used to identify the substance and to verify that the person tested probably used the substance. There is general agreement in the scientific community that urine concentrations for most substances cannot be correlated with performance, health, or safety, whereas no generally accepted scientific foundations correlate urine concentrations with performance (43). Documentation does not exist that could serve as a basis for predicting the quantity, strength, frequency of use, or some prior urine concentration for most drugs.

Some have attempted to interpret the meaning of urine cannabinoid concentrations. Ellis et al. (44) detected cannabinoids in urine from 4 to 34 days after last use in light marijuana users (those who use it once weekly or less), and from 1 to 81 days later in heavy users (those whose use is daily or more often), depending on whether a 20- or 100-μg/L cutoff immunoassay was used. Sometimes the urine concentration dropped below the cutoff value but was positive in a later urine specimen, with no evidence of intervening drug use. A subject whose urinary cannabinoid concentration dropped from 2700 μg/L to about 150 μg/L in 23 days showed no signs of impairment during this period (45).

Hollister et al. (46) found it is unlikely that a range of plasma THC concentrations could be reliably equated with impaired performance.

McBurney, Bobbie, and Sepp (47) determined that recent smoking of marijuana, that is, within 6 hr, was indicated by THC concentrations in plasma in excess of 2–3 μg/L and 8β,11-dihydroxytetrahydrocannabinol concentrations in urine in excess of 15–20 μg/L. Their report was based on acute exposure in 10 subjects to about 10 mg THC. Tests for 8β,11-dihydroxytetrahydrocannabinol are not currently available from commercial laboratories.

Urines containing 637.7 and 847 μg/L of THC-COOH were obtained from two dead victims of motor vehicle crashes when no THC, THC-COOH, or hydroxy-THC (THC-OH) was found in their bloods (48).

A preliminary report stated that there were performance decrements on a flight simulator by pilots 24 hr after smoking strong marijuana cigarettes (49), but urines and bloods were not tested for cannabinoids, alcohol, or for any other drugs, and the study lacked adequate controls. Another preliminary report dealt with a similar study of pilots after they had drunk enough ethanol for their blood concentrations to reach 100 mg/dL (0.10%). Significant decrements in performances were still evident when they were tested 14 hr later when their bloods contained no ethanol (50). One might conclude from these two reports that a pilot whose blood contained no alcohol and no THC but did contain THC-COOH could have a performance decrement because the blood ethanol concentration could have been 100 mg/dL about 14 hr earlier and(or) marijuana might have been smoked 24 hr before the time of the blood test. Or the pilot might not be impaired by either drug.

A recent review of the subject of drugs and driving offers evidence of decrements in some tasks after using drugs that may be relatable to driving (51), but the only drug concentrations reported are plasma concentrations of diazepam. A "Guide for Presumptive Indication of Intoxication or Being Under the Influence of Alcohol and Drugs" (52) presents concentrations for some drugs in blood and urine that presumably reflect impairment. The naiveté of the author is demonstrated by listing blood concentrations and is emphasized by listing urine concentrations. No foundation is presented for the guesses. The U.S. government has prepared guidelines for the performance of urine drug testing (53) and for the certification of laboratories doing such testing (54). The concentrations recommended as cutoffs for the screening and confirmatory methods are based on analytical capability and not on any possible relationship to health, safety, or performance.

Experts who correlate drug concentrations with "under the influence," intoxication, impaired or improved performance, or effects on health or safety should give documentation of the scientific foundation for their opinions. Except for alcohol there is a lack of data in the literature on blood concentration and on drugs in urine upon which an expert may base an opinion of drug-related impairment or improvement. Calculating the concentrations of ethanol in blood to some time prior to the test presents problems, and back-extrapolation of the concentrations of other drugs is rarely possible. Marijuana and cocaine are good illustrations of this problem. The blood concentrations of the parent drugs peak very rapidly, then decrease very rapidly, then decrease at varying rates for long periods of time. The drug metabolites then appear and their concentrations decrease more slowly. Attempts to measure the concentrations of the parent drug and the metabolite would be useful if chronic users did not have a background concentration of metabolite from prior use.

8. DOCUMENTATION

Collectors of urine specimens of workers should start a chain of custody that should be continued by everyone who handles the specimen until the urine is expended or discarded. The original specimen should be placed into two containers. One container should be sent for analysis, and the other should be sealed and preserved by refrigerating or freezing. When the results of the original specimen are challenged, the second specimen should also be divided. One portion should be retested and the other given to the accused user for independent testing. This procedure could eliminate chain-of-custody arguments.

Laboratories testing urines of workers should be prepared to document fully everything that happened to the specimens, how the results were obtained, and what the results mean. "Documentation of all aspects of the testing process must be available. This documentation will be maintained for at least 2 years and will include: personnel files on analysts, supervisors, and directors, and all individuals authorized to have access to specimens; chain of custody documents; quality assurance/quality control records; all test data; reports; performance records of proficiency testing; performance on accreditation inspections; and hard copies of computer-generated data" (53).

9. CONCLUDING REMARKS

The need for testing workers' urines has not been established but if random testing is done, the choice of drugs to be detected should be carefully made, and quality testing and supportable interpretations should be available. The particular screening methods chosen and any possible interferences are not important if the confirmatory method used produces adequate individual characteristics for a "most probable" identification. There is general agreement in the scientific community that properly conducted GC/MS methods can meet these criteria. Even if a drug or metabolite in urine is positively identified and precisely quantified, there is no scientific basis for forming opinions as to when, how often, and how much drug was used or as to the past, present, or future effect of the drug on the performance, health, or safety of the worker. Any worker whose behavior, performance, health, or safety changes without known cause should be given a physical examination, including tests of blood and urine for the usual clinical chemistry, for drugs and metabolites, and for chemicals, radiation, and other hazards that the worker might be exposed to in the workplace.

REFERENCES

1. G. D. Lundberg, *J. Am. Med. Assoc.*, *256*, 3003–3005 (1986).

2. A. J. McBay, *J. Am. Med. Assoc.*, *255*, 40 (1986).

3. A. J. McBay, *Alcohol Drugs Driving*, *2*, 51–59 (1986).

4. *Army Times*, Nov. 11, 1985, p. 1.

5. A. J. McBay and R. P. Hudson, *J. Forensic Sci.*, *32*, 575–579 (1987).

6. "Guidelines for Data Acquisition and Data Quality Evaluation in Environmental Chemistry," *Anal. Chem.*, *52*, 2242–2249 (1980).

7. "EMIT d.a.u. Cannabinoid 20 Assay," SYVA, Palo Alto, CA, Sept. 1984.

8. R. H. Schwartz, G. F. Hayden, and M. Riddle, *Am. J. Dis. Child.*, *139*, 1093–1096 (1985).

9. A. P. Mason, M. Perez-Reyes, A. J. McBay, and R. L. Foltz, *J. Anal. Toxicol.*, *7*, 172–174 (1983).

10. E. J. Cone and R. E. Johnson, *Clin. Pharmacol. Ther.*, *40*, 247–256 (1986).

11. B. Law, P. A. Mason, A. C. Moffat, R. I. Gleadle, and L. J. King, *J. Pharm. Pharmacol.*, *36*, 289–294 (1984).

12. D. Lorenzen, *SYVA Lett.*, Feb, March, April, June, and July 1986.

13. K. Bjerver, J. Jonsson, A. Nilsson, J. Schuberth, and J. Schuberth, *J. Pharm. Pharmacol.*, *34*, 798–801 (1982).

14. M. A. ElSohly, D. F. Stanford, and H. N. ElSohly, *J. Anal. Toxicol., 10,* 256 (1986).

15. R. K. Siegel, M. A. ElSohly, T. Plowman, P. M. Rury, and R. T. Jones, *J. Am. Med. Assoc., 255,* 40 (1986).

16. A. J. McBay, *Lab. Management, 23,* 36–43 (1985).

17. B. Law, P. A. Mason, A. C. Moffat, and L. J. King, *J. Anal. Toxicol., 8,* 19–22 (1984).

18. J. J. Tassett, T. J. Schroeder, and A. J. Pesce, *J. Anal. Toxicol., 10,* 258 (1986).

19. R. P. Betah, *Tox. Talk, 11,* 3 (1987).

20. "KDI QUIK TEST Drug Screen," fact sheet (and personal communication), Keystone Diagnostics, Columbia, MD., Dec. 1986.

21. "Low-cost Kit for Drug Screening now Available," *Chem. Eng. News, 64,* 5 (1986).

22. "The Melanin Defense, Debated by Woodford and McBay," *Subst. Abuse Rep., 17,* 3–8 (1986).

23. Frye v. United States 293 F. 1013 (D.C. Cir. 1923).

24. Fed. R. Evidence 401,403,702.

25. C. E. Leal, Wake Forest Law Rev., *20,* 391–412 (1984).

26. H. J. Hansen, S. P. Caudill, and J. Boone, *J. Am. Med. Assoc., 253,* 2382–2387 (1985).

27. R. E. Willette, in R. L. Hawks and C. N. Chang, Eds., NIDA Research Monograph 73, National Institute on Drug Abuse, Rockville, MD, 1986, p. 22.

28. K. M. Dubowski, *Alcohol, Drugs Driving, 2,* 13–46 (1986).

29. M. F. Mason, *J. Anal. Toxicol., 5,* 201–208 (1981).

30. M. L. Abercrombie and J. S. Jewell, *J. Anal. Toxicol., 10,* 178–180 (1986).

31. J. L. Frederick and J. Green, *J. Anal. Toxicol., 9,* 116–120 (1985).

32. C. A. Sutheimer, R. Yarborough, B. R. Hepler, and I. Sunshine, *J. Anal. Toxicol., 9,* 156–160 (1985).

33. M. J. Kogan, J. Al Razi, D. J. Pierson, and N. J. Willson, *J. Forensic Sci., 31,* 494–500 (1986).

34. J. Irving, B. Leeb, R. L. Foltz, C. E. Cook, J. Bursey, and R. E. Willette, *J. Anal. Toxicol., 8,* 192–196 (1984).

35. M. A. Peat, M. E. Deyman, and J. R. Johnson, *J. Forensic Sci., 29,* 110–119 (1984).

36. NCAA Drug Testing Program, NCAA Publishing, Mission, KS, 1986–1987.

37. *Durham Morning Herald* (AP), Sept. 25, 1986, p. 1D.

38. O. R. Idowu and B. Caddy, *J. Forensic Sci. Soc., 22,* 123–135 (1982).

39. S. J. Gross, T. E. Worthy, L. Nerder, E. G. Zimmermann, J. R. Soares, and P. Lomax, *J. Anal. Toxicol., 9,* 1–5 (1985).

40. L. K. Thompson, D. Yousefnejad, K. Kumor, M. Sherer, and E. J. Cone, *J. Anal. Toxicol., 11,* 36–38 (1987).

41. L. E. Norton and J. C. Garriott, *Am. J. Forensic Med. Pathol., 4,* 185–188 (1983).

42. W. A. Baumgartner, "Preliminary proposal: Employee drug screening by hair analysis," IANUS Foundation, Los Angeles, 1986, pp. 1–4.

43. R. V. Blanke, Y. H. Caplan, R. T. Chamberlain, et al., *J. Am. Med. Assoc., 254,* 2618–2621 (1985).

44. G. M. Ellis, M. A. Mann, B. A. Judson, N. T. Schramm, and A. Tashchian, *Clin. Pharmacol. Ther., 38,* 572–578 (1985).

45. R. C. Baselt, *J. Anal. Toxicol., 8,* 16A (1984).

46. L. E. Hollister, H. K. Gillespie, A. Ohlsson, J-E. Lindgren, A. Wahlen, and S. Agurell, *J. Clin. Pharmacol., 21,* 171S–177S (1981).

47. L. J. McBurney, B. A. Bobbie, and L. A. Sepp, *J. Anal. Toxicol., 10,* 56–64 (1986).

48. J. C. Garriott, V. J. M. DiMaio, and R. G. Rodriguez, *J. Forensic Sci., 31,* 1274–1282 (1986).

49. J. A. Yesavage, V. O. Leirer, M. Denari, and L. E. Hollister, *Am. J. Psychiatry, 142,* 1325–1329 (1985).

50. J. A. Yesavage and V. O. Leirer, *Am. J. Psychiatry, 143,* 1546–1550 (1986).

51. H. Moskowitz, *Accid. Anal. Prev., 17,* 281–345 (1985).

52. R. E. Willette, "Testing Employees for Alcohol and Drugs," Bureau of National Affairs Conference, Washington, DC, Nov. 8, 1984.

53. "Mandatory Guidelines for Federal Workplace Drug Testing Programs," *Federal Register, 53,* 11970–11984 (1988).

54. "Scientific and Technical Guidelines for Federal Drug Testing Programs; Standards for Certification of Laboratories Engaged in Urine Drug Screening for Federal Agencies, Appendix A—Proficiency Test Program," *Federal Register, 52,* 30647–30652 (1987).

INDEX

295

(*continued from front*)